O. S. Fowler

The Practical Phrenologist and Recorder and Delineator of the Character and Talents ...

A Compendium of Phreno-Organice Science

O. S. Fowler

The Practical Phrenologist and Recorder and Delineator of the Character and Talents ...
A Compendium of Phreno-Organice Science

ISBN/EAN: 9783337156121

Printed in Europe, USA, Canada, Australia, Japan

Cover: Foto ©berggeist007 / pixelio.de

More available books at **www.hansebooks.com**

THE
PRACTICAL PHRENOLOGIST;

AND

RECORDER AND DELINEATOR

OF THE

CHARACTER AND TALENTS

OF

As marked by

A COMPENDIUM

OF

PHRENO–ORGANIC SCIENCE.

BY

O. S. FOWLER,

PRACTICAL PHRENOLOGIST, LECTURER, FORMER EDITOR OF "AMERICAN PHRENOLOGICAL JOURN
AND AUTHOR OF "FOWLER ON PHRENOLOGY," "FOWLER ON PHYSIOLOGY," "SELF-CULTURE
"MEMORY," "RELIGION," "MATRIMONY," "HEREDITARY DESCENT," "LOVE AND
PARENTAGE," "MATERNITY," "AMATIVENESS," "SELF INSTRUCTOR," "HOME
FOR ALL," "ANSWER TO HAMILTON," "VINDEX," ETC., ETC., ETC.

BOSTON:
O. S. FOWLER, 514 TREMONT STREET.

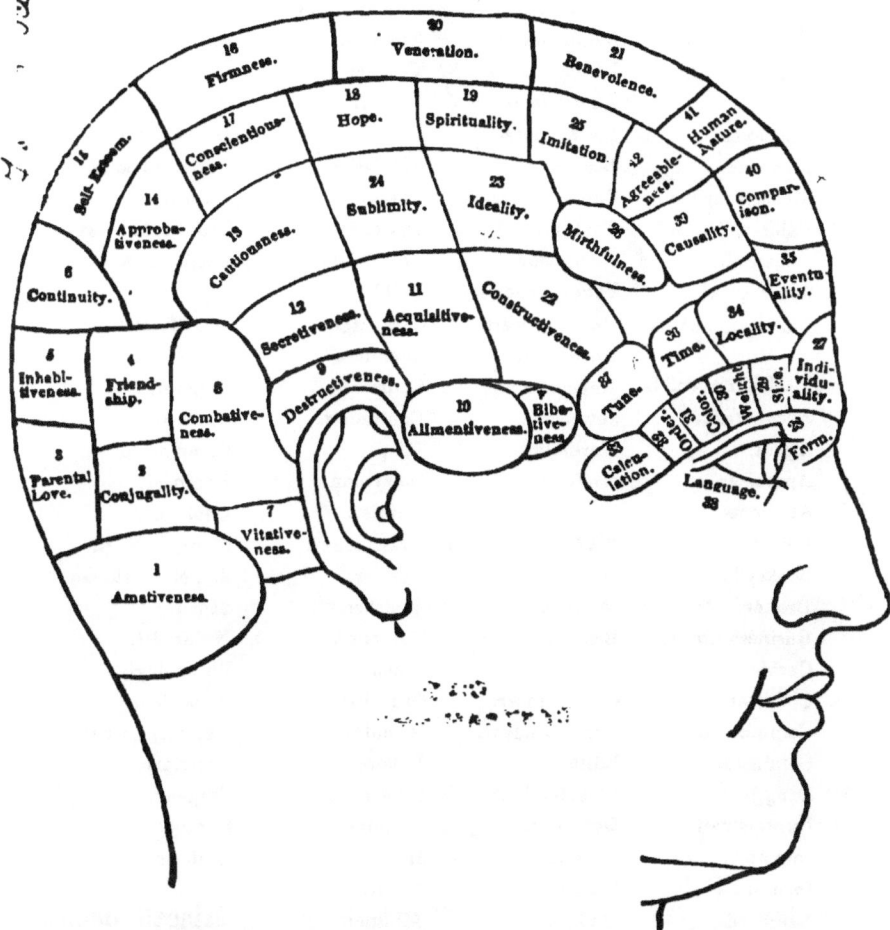

NAMES, NUMBERING, AND DEFINITIONS OF THE FACULTIES.

1. AMATIVENESS. — Sexual love, fondness, passion.
2. CONJUGALITY. — The pairing instinct, one love.
3. PARENTAL LOVE. — Care for offspring, and young.
4. FRIENDSHIP. — Sociability, clinging to friends.
5. INHABITIVENESS. — Love of home, patriotism.
6. CONTINUITY — Application, finishing, continuing.
7. VITATIVENESS. — Clinging to life, resisting disease.
8. COMBATIVENESS. — Defense, courage; force, etc.
9. DESTRUCTIVENESS. — Executiveness, severity.
10. ALIMENTIVENESS. — Appetite, relish, greediness.
11. ACQUISITIVENESS. — Frugality, saving, industry.
12. SECRETIVENESS. — Self-control, policy, art, tact.
13. CAUTIOUSNESS. — Guardedness, safety, prudence.
14. APPROBATIVENESS. — Pride of character, honor.
15. SELF-ESTEEM. — Self-respect, dignity, authority.
16. FIRMNESS. — Stability, perseverance, willfulness.
17. CONSCIENTIOUSNESS. — Duty, right, truth, justice.
18. HOPE. — Expectation. anticipation, enterprise.
19. SPIRITUALITY. — Intuition, prescience, faith.
20. VENERATION — Worship, adoration, obedience.
21. BENEVOLENCE — Sympathy, kindness goodness.
22. CONSTRUCTIVENESS. — Ingenuity, invention.
23. IDEALITY. — *Taste*, love of beauty, poetry
24. SUBLIMITY. — Love of grandeur, vastness, etc
25. IMITATION. — Copying, aptitude, mimickry.
26. MIRTH. — Fun, wit, ridicule, facetiousness
27. INDIVIDUALITY. — Observation, desire to *see*.
28. FORM. — Memory of *shape*, looks, persons.
29. SIZE. — Measurement of quantity, distance
30. WEIGHT. — Control of motion, balancing.
31. COLOR. — Discernment and love of colors.
32. ORDER. — *Method*, system, doing by *rule*.
33. CALCULATION. — Mental arithmetic, reckoning
34. LOCALITY. — Memory of place, position, etc.
35. EVENTUALITY. — Memory of facts, events, etc.
36. TIME. — Telling *when*, time of day, dates, etc
37. TUNE. — Musical love, ecstasy, and talent
38. LANGUAGE. — *Expression by* words, acts, etc.
39. CAUSALITY. — *Planning*, thinking, reason, sense
40. COMPARISON. — Analysis, inferring, critic.
41. HUMAN NATURE. — Perception of character.
42. SUAVITY — *Pleasantness* blandness, blarney

BUSINESS ADAPTATIONS

IN A SCALE FROM 1 TO 7.

Artistical.
Architect.
Designer.
Engraver.
Musician.
Music Teacher.
Painter, Orna-
　mental.
do., Portra ∵.
Photographe ∵.

Commercial.
Accountant.
Agent.
Appraiser.
Auctioneer.
Banker.
Bookseller.
Broker.
Business Corres.
Cashier.
Collector.
Commis. Mer.
Conductor.
Druggist.
Expressman.
Importer.
Insurance.
Landlord.
Merchant.
　Principal.
Publisher
Salesman.
Shipping Clerk.
Speculator.
do., Real Estate.
Superintendent.
Trader.

　MARRY one

Retail Dealer.
Wholesale do.
Dealer in —
Boots, shoes.
Leather.
Cattle, horses.
Coal, lumber.
Dry-goods.
Fancy Articles.
Grain, groceries.
Hardware.
Implements.
Jewelry.
Marketing.
Useful Articles.

Professional.
Actor.
Author.
Bishop.
Clergyman.
Conveyancer.
Correspondent.
Editor.
Elocutionist.
Governor.
Governess.
Historian.
Judge.
Lawyer.
Lecturer.
Literature.
Linguist.
Officer.
Poet.
Politician.
Professor.
Proof-reader.

In Size 6

Reporter.
Teacher
Writer.

Mechanical.
Baker.
Blacksmith.
Boss Workman.
Builder.
Carpenter.
Chandler.
Compositor.
Contractor.
Cooper.
Dairyman.
Dentist.
Dressmaker.
Farmer.
Finisher.
Gardener.
Gunsmith.
Gas Fitter.
Inventor.
Laborer.
Locksmith.
Machinist.
Mason.
Miller.
Milliner.
Paperer.
Plumber.
Printer.
Tailor.
Tanner.
Tinsmith.
Turner.
Seamstress.
Stonecutter.

Height 6

Shipbuilder.
Upholsterer.
Manufacturer of—
Boots, shoes.
Fancy Articles.
Furniture.
Trunks, harness.
Useful Articles.

Scientific.
Anatomist.
Captain.
Chemist.
Commander.
Engineer.
Geologist.
Manager R. R.
do., of Workmen.
Miner.
Naturalist.
Phrenologist.
Physician.
Representative.
Secretary.
Surgeon.
Surveyor.
Statesman.

Miscellaneous.
Fisherman.
Housekeeper.
Livery Keeper.
Matron.
Nurse.
Restaurant.
Teamster.
Waiter.
Watchman.

Complexion *dui*

CONDITIONS.	7 Very Large.	6 Large.	5 Full.	4 Average.	3 Moderate.	2 Small.	Cultivate.	Restrain.	Marry one having
Inches. 2 0 7 8	PAGE								
Size of Brain,	6	7	7	7	8	8			
Organic Quality........	16	17	17	17	17	17	17	18	
Health..............	21	21	21	21	22	22	22	22	6
Vital Power,	25	25	26	26	26	26	26	27	6
Breathing Power......	28	28	29	29	29	29	29		
Circulatory Power....	30	30	30	30	30	30	30	31	C
Digestive Power.......	31	31	31	31	31	32	32	33	6
Motive Power,	34	35	35	36	36	37	38	38	
Mental Power,	39	40	41	41	41	41	87	47	
Activity..............	46	46	46	46	46	47	47	47	
Excitability	47	48	48	48	48	48	48	48	
DOMESTIC GROUP.	66	66	66	66	66	65			
1. Amativeness........	67	68	69	69	70	70	71	71	8
2. Conjugality	72	72	72	72	73	73	73	73	C
3. Parental Love.......	74	74	75	85	75	75	76	76	
4. Friendship	76	77	77	78	78	78	78	78	
5. Inhabitiveness......	79	79	79	80	80	80	80	80	
6. Continuity.........	80	80	81	81	81	81	82	82	
SELFISH GROUP.	82	82	83	83	83	83	83	83	
7. Vitativeness	84	84	84	84	84	84	84	84	
8. Combativeness......	85	85	86	86	87	87	87	87	
9. Destructiveness	88	88	89	89	89	89	90	90	
10. Alimentiveness......	91	91	91	91	91	91	92	92	
11. Acquisitiveness	94	94	95	95	95	95	96	96	
12. Secretiveness........	97	97	97	98	98	98	99	99	
13. Cautiousness........	99	100	101	101	101	101	102	102	
14. Approbativeness.....	102	103	103	103	104	104	104	104	
15. Self-Esteem.........	105	105	106	106	106	107	107	107	
16. Firmness...........	108	108	109	109	109	109	109	109	
MORAL GROUP.	110	110	110	111	111	111	111	111	

CONDITIONS.	7 Very Large.	6 Large.	5 Full.	4 Average.	3 Moderate.	2 Small.	Cultivate.	Restrain.	Marry one having
17. Conscientious.......	112	112	113	113	114	114	114	114	
18. Hope	115	115	116	116	116	117	117	117	
19. Spirituality	117	118	118	118	118	118	118	119	
20. Veneration	119	120	121	121	121	121	121	121	
21. Benevolence	122	123	123	123	123	123	123	124	
SELF-PERFECTIVES.	124	124	124	124	124	125	125	125	
22. Constructiveness	125	125	126	126	126	126	126	127	
23. Ideality	127	127	128	128	128	128	129	129	
24. Sublimity	129	129	130	130	130	130	130	130	
25. Imitation	130	131	132	132	132	132	132	132	
26. Mirthfulness.......	133	133	134	134	134	134	134	134	6
INTELLECTUALS.	135	135	135	135	135	135	135	136	
PERCEPTIVES.	136	136	136	137	137	137	137	137	
27. Individuality.......	137	138	138	139	139	139	139	139	
28. Form	139	139	140	140	140	140	140		
29. Size...............	141	141	141	141	141	141	141	142	
30. Weight........... .	142	142	142	142	143	143	143	143	
31. Color..............	143	143	144	144	144	144	144	144	
32. Order............	144	145	145	145	145	145	146	146	
33. Calculation........	146	147	147	147	147	147	147	147	
34. Locality...........	147	147	148	148	148	148	148	148	
LITERARY FACULTIES.	148	148	149	149	149	149	149	149	
35. Eventuality........	149	149	150	150	150	151	151	151	
36. Time..............	151	151	151	151	152	152	152	152	
37. Tune..............	152	152	152	153	153	153	153	153	
38. Language	153	154	154	155	155	155	155	156	
REFLECTIVES.	156	156	156	156	156	156	156	156	
39. Causality..........	157	157	158	158	158	158	158	159	
40. Comparison........	159	159	160	160	160	160	160	160	
41. Human Nature	161	161	161	161	161	161	161	162	
42. Agreeableness.......	162	162	162	162	162	162	162	162	

PREFACE.

To TEACH LEARNERS those organic conditions which indicate character, is the first object of this manual. And to render it accessible to all, it condenses facts and conditions, rather than elaborates arguments, — because to expound Phrenology is its highest proof, — states laws and results, and leaves them upon their naked merits; embodies recent discoveries, and crowds into the fewest words and pages just what learners most need to know, and hence requires to be STUDIED rather than merely read. "Short, yet clear," is its motto. Its analysis of the faculties and numerous engravings embody the results of observation and experience.

To RECORD CHARACTER is its second object. In doing this, it describes those organic conditions which affect and indicate character in SEVEN degrees of power — very large, large, full, average, moderate, small, and very small — indicated by the seven numerals 7, 6, 5, 4, 3, 2, and 1.

Those organs and conditions marked 7, or very large, are sovereign in their influence over character and conduct, and, combining with those marked large, direct and control the feelings and actions. Those marked 6, or large, have a powerful and almost controlling influence, both singly, and especially in combination, and press the smaller ones into their service. Those marked 5, or full, play subordinate parts, yet their influence is considerable, though more potential than apparent. Those marked 4, or average, have only a medium influence, and mainly in combination with larger ones. Those marked 3, or moderate, are below par in fact, and still more so in appearance; exert but a subordinate influence; and leave character defective in these respects. Those marked 2, or small, are so deficient as easily to be perceived; leave their possessor weak and faulty in these respects, and should be assiduously cultivated; while those marked 1 are very small, and render their possessor almost idiotic in these respects.

Those who have their physiological and phrenological conditions correctly marked in the accompanying table, are referred in it to those paragraphs in this and the Author's other works which both describe themselves, and also contain specific directions how to *perfect* their characters, and improve children. Its plan for recording character is seen at a glance in the following —

EXPLANATION OF THE TABLES.

The examiner will mark the power, absolute and relative, of each function and faculty, by placing a figure, dot, or dash on a line with the name of the organ marked, and in the column headed "large" or "small," according to the size of the organ marked, while the printed figure in the square thus marked refers to those pages in this book where, under the head "large," "small," etc., will be found a description of the character of the one examined in respect to that faculty, and at the end of this description, in the book, another figure will be found, which refers to FOWLER'S "PHRENOLOGY," a standard work, in which will be found an extended description of those shadings of character caused by various combi- nations of faculties, while in the two right-hand columns but one, in the columns headed "cultivate" and "restrain," are figures referring to pages in this work where directions for cultivating and restraining may be found; and at the close of these sentences are figures which refer the reader to the *numbered paragraphs* in three books, entitled "Physiology," "Self-culture," and "Memory," called, when bound together, "Education Complete," where will be found extended directions for self-improvement and the management of children. For example:

CONDITIONS.	7 Very Large.	6 Large.	5 Full.	4 Aver- age.	3 Moder- ate.	2 Small.	Culti- vate.	Re- strain	Marry one having
15. Conscientious- ness	PAGE — 112	112	113	113	. 114	114	. 114	— 117	
16. Hope.........	. 115	115	116	116	— 116	117	— 114	. 117	— 6

This section of the table presupposes that two persons, A and B, have been marked upon it, A with a dash, B with a dot, and shows that A's Conscien- tiousness is very large, and that his character is described on page 112, under the head "very large," and that it should be restrained, which he is told how to do in "Education," numbered paragraph 268, under the head "restrain," but that Hope is moderate, which is described under "moderate" on page 116, and requires to be cultivated, which he is told how to do in "Education," num- bered paragraph 272, under "cultivate;" but that B's Conscientiousness is moderate, and is described on page 114, under "moderate," and to be cultiva- ted, and is shown how under "cultivate," in paragraph 268 of "Education," but that B's Hope is very large, and is described on page 115, under "very large," and is to be restrained, see "Education," paragraph 273, under "re- strain," and must marry one having Hope large

The right-hand column, headed "marry one having" shows to what tempera- ments and phrenological developments the one marked is best adapted. See the Author's work on "The Family." The points left unmarked are less material, concerning which choose according to your own tastes.

Several persons can be marked on one table by using a dot for one, and dashes, horizontal, perpendicular, slanting to the right, left, etc., or different colored pencils, for each of the others, so that all the members of a family, or a group of friends, can thus be marked on one table, or all transferred from each table to that of all the other tables, so that each can have the records and descriptions of all the others, and all of each other.

When an organ is about half-way between two sizes, it is represented by two dashes or dots, as 5 to 6, or 3 to 4, etc., which is equivalent to $5\frac{1}{2}$ or $3\frac{1}{2}$. In these cases both sentences referred to may be read, and a medium between the two will be appropriate.

The sign +, *plus*, signifies about one third of a degree more, and —, *minus*, one third of a degree less, than the marks indicate; thus giving virtually a scale of twenty-one degrees.

PRACTICAL PHRENOLOGIST, ETC.

DEFINITION AND PROOFS OF PHRENOLOGY.

PHRENOLOGY — derived from the two Greek words, φρήν, mind, and λόγος, discourse — points out certain cause and effect relations existing between particular FORMS or developments and conditions of the brain, and their accompanying MANIFES-TATIONS of the mind, and predicates the respective peculiari-ties of the character and talents of different persons from the forms, sizes, and other organic conditions of their brains.

It must, of necessity, be either true or false. If true, it constitutes a distinctive department of Nature, and must, therefore, harmonize with all her other departments ; but, if false, it must needs conflict with her laws and facts. Surely, then, it need not be difficult to ascertain whether it thus agrees or conflicts with Nature.

Its distinctive doctrines are that : —

I. The mind is composed of different PRIMARY POWERS or forces, called faculties, each of which manifests a specific CLASS of the mental functions.

Thus the feeling of sympathy is the product of one mental faculty, called Benevolence, and memory of facts is put forth by another called Eventuality ; while reasoning by induction is exercised by another, called Comparison, and thus that there exist as many primary mental capacities as man can experience different kinds of mental functions.

That the mind consists of several elemental faculties, and not of one single power, is evinced by —

1. The different inherent traits and instincts of different persons and animals. The duck " takes to " water, and eagle to crags: the lion to flesh, and horse to grain ; one man to letters and another to me-

chanics ; one to philanthropy and another to money, etc., because impelled thereto by strong innate proclivities. " Poets are *born*, not educated."

2. Monomania is consequent on the derangement of one mental faculty, while all the other faculties are sane. If it consisted in only one element, all its functions must needs be deranged or sane *together*, according as this one was sane or insane.

3. If all kinds of talent originated from this one element, it must be *equally* strong or weak in everything, whereas, instead, one man is often great in one or two respects yet deficient in others, like Blind Tom, a musical and a mimicking prodigy, though a natural fool — some remembering some things but forgetting others ; some great mechanics but poor speakers. And thus of most men in a greater or less degree.

4. If the mind consisted of but one element, it could do but one thing at the same instant, and must, of course, stop all previous functions the instant it commenced any and every new one ; must stop remembering the instant it began to think, and stop thinking the instant it began to remember, and suspend both and all its other functions the instant it began to talk. Yet, in that case, how could it talk at all, for how could it remember what it would say, or say anything while it remembered ? This doctrine of the oneness of the mental faculty is contradicted, while that of its plurality is proved, by every single mental fact and phenomenon bearing on this point.

Therefore the mind must necessarily be composed of just as many separate faculties as it can put forth distinct classes of operations — a primary faculty for each class.

And these " special geniuses " are caused by one faculty being strong, whilst another is weak — that is, by their different degrees of power, in different persons and modes of action. A mental faculty is : —

1. That which puts forth a *distinct class* or kind of mental function.

2. That which appears or disappears *earlier or later* in life than others.

3. That which can act or rest, be healthy or diseased, strong or weak, *independently* of the others.

4. That which is propagated *separately*, and in *different degrees* of power.

These faculties, so *embodied* that they act collectively, are the only instrumentalities of all we feel, do, and are, and collectively constitute our consciousness, selfhood, personality, and life-entity.

II. EACH FACULTY is exercised by means of a particular POR-TION of the brain, called its organ.

Proofs : —

1. All functions whatsoever, are always put forth by means of OR-GANS, never without them. Not one function anywhere in Nature but is exercised through some *organ*. That is, organism is Nature's only means of functionism.

2. Every class of functions is manifested only through its own specific organ, created expressly therefor. Thus, who can ever see except through eyes, or hear without ears, or move without muscles, or fulfill any function whatsoever, except in and by Nature's specific organs, expressly adapted thereto ? And she always employs one, fifty, or five hundred organs, whenever she has one, fifty, or five hundred functions — its own particular organ for each class of functions. Indeed, what is Nature's sole rationale and end of all matter, throughout all its forms, but to furnish the *organisms* requisite for executing her respective functions ?

Of course this organic institute of Nature, so indispensable throughout all her other departments, is equally useful and necessary in each of her mental functions. Each faculty of the mind must, therefore, have its own specific organ, through which alone it can be exercised.

III. THE BRAIN IS THE ORGAN OF THE MIND.

This doctrine is universally admitted. Its proofs are : —

1. It was not made for nought, but was created to execute *some* function.

2. Its structure, and everything appertaining to it, show that it fulfils altogether the most important function of man, which is, of course, the mental.

3. Anatomy proves that it exercises a PART and therefore ALL of the mental operations.

4. Every natural fact which bears on this point proves it. Not one militates against it.

IV. THE BRAIN IS A CLUSTER OF ORGANS, each expressing one faculty only.

1. ANATOMY proves that different parts of the brain perform different mental functions: that sight is executed by one portion, and hearing, tasting, etc., each by others ; therefore every other mental power must have its own specific cerebral organ.

2. INSANITY is caused by inflammation of the brain. This is proved by those mental derangements consequent on many fevers ; by delirium tremens ; by certain injuries of the brain, impairing specific mental powers ; by a softening of the brain, weakening the mentality ; and many similar ranges of facts.

3. MONOMANIA, or insanity on one subject, coexisting with sanity on all others — the usual form of mental derangement — is obviously caused by disorder in one of these cerebral organs, whilst the others are sound.

4. INJURIES OF THE BRAIN furnish still more demonstrative proof. If Phrenology is true, to inflame Tune, for example, would create a singing disposition ; Veneration, a praying desire ; Cautiousness, groundless fears ; and so of all the other organs. And thus it is. Nor can this class of facts be evaded. They abound in all phrenological works, especially periodicals, and drive and clench the nail of proof.

V. PARTICULAR CHARACTERISTICS are always accompanied and indicated each by its own SPECIFIC FORM.

1. Nature classifies all her productions into orders, genera, species, etc., and annexes specific forms to each, so that the same forms always accompany and indicate the same traits of character. Thus one form of tree and leaf always accompanies and indicates oak characteristic, another peach, and thus of the shape of every tree, vegetable, and thing that grows. Botany is based on this principle, and consists in its various ramifications.

2. Every branch of Natural History furnishes infinite ranges of illustrations of this same law. Every kind of fish, fowl, and creeping thing, from the beginning of time, always has kept, and will keep, its own specific configuration, to which each individual of every class, genius, and species, conform. Thus all dogs have one general form, all cats another, all bovines still another, and thus of all elephants, all humans, all that lives.

So, too, all bull-dogs have one variation of this canine form, all greyhounds another, all spaniels still another, and thus of all other varieties of dogs, cats, fish, fowls, insects, races of men — everything.

3. Any anatomist too, can predicate, with infallible certainty, just from the mere shape of the smallest bone of any unknown animal or human being, its natural history, and all about its detailed characteristics and instincts. Then since the size and shape of every leaf, scale, feather, bone, etc., of every living creature, vegetable, mineral, and

thing, tell us infallibly all about its specific characteristics, of course all the various forms of the head must also indicate and accompany, each its own specific mental traits. Shall universal form proclaim universal character, and shall not specific head-shapes also proclaim particular mental dispositions and talents?

4. In and by the very nature and constitution of things, specific forms are linked each to its particular mental speciality. Therefore, every distinct form of the brain and head indicates some particular proclivity or passion. This is but one phase of a universal ordinance of all things.

VI. SIZE, other things being equal, indicates the POWER of function.

That this proposition expresses a general law, is evinced by the general fact that the larger the pieces of wood, iron, etc., are, the stronger they are; that larger horses, persons, etc., are proportionally more powerful than smaller, and thus of everything else. Though sometimes smaller men, horses, etc., are stronger, can lift, draw, and endure more than others that are larger, because they are different in organic quality, health, etc., yet where the *quality* is the same, whichever is largest is proportionally the most powerful.

And this undisputed law of things is equally true of the brain, and that mental power put forth thereby. All really *great* men have great heads — merely smart ones, or those great only in certain faculties or specialities of character, not always. The brains of Cuvier, Byron, and Spurzheim were among the very heaviest ever weighed. True, Byron's *hat* was small, doubtless because his brain was conical, and most developed in its base; but its great *weight* establishes its great size. So does that of Bonaparte. Besides, he wore a very large hat — one which passed clear over the head of Colonel Lehmenouski, one of his body-guard, whose head measured 23½ inches, so that Bonaparte's head must have measured nearly or quite 24 inches. Webster's head was massive, measuring over 24 inches, and Clay's 23½; and this is about Van Buren's size. Chief Justice Gibson's, the greatest jurist of Pennsylvania, was 24¼; and Hamilton's hat passed over the head of a man whose head measured 23½. Burke's head was immense, so was Jefferson's, while Franklin's hat passed over the ears of a 24 inch head. Judge McLean's head exceeded 23½ inches. The heads of Washington, Adams, and a thousand other celebrities, were also very large. Bright, apt, smart, literary, knowing, even eloquent men, etc., often

have only average, even moderate-sized heads, because endowed with
the very highest organic *quality*, yet such are more admired than commanding ; more brilliant than powerful ; more acute than profound.
Though they may show off well in an ordinary sphere, yet they are not
the men for *great* occasions ; nor have they that giant *force* of intellect which molds and sways nations and ages. The phrenological law
is, that size, *other things being equal*, is a measure of power ; yet these
other conditions, such as activity, power, motive, health, physiological
habits, etc., increase or diminish the mentality even more than size.
Quality is *more* important than quantity, but true greatness require*
both cerebral quantity and quality.

Still, those again who have very large heads, are sometimes dull,
almost foolish, because their organic quality is low. As far, then, as
concerns Phrenology itself, this doctrine of size appertains to the different organs in the *same head*, rather than to different heads. Still
this doctrine, that size is the measure of power, is no more a special
doctrine of Phrenology than of every other department of nature. And
those who object to this science on this ground are objecting to a known
law of things. If size were the *only* condition of power, their cavils
might be worthy of notice ; as it is, they are not.

Though tape measurements, taken around the head, from Individuality to Philoprogenitiveness or Parental Love, give some idea of the
size of the brain; the fact that some heads are round and others long,
some low and others high, etc., so modifies these measurements that
they do not convey any very correct idea of the actual quantity of brain.
Yet these measurements range somewhat as follows in adults : —

7, or Very Large, 23¾ inches, and upward ; 6, or Large, from 22¾
to 23¾ ; 5, or Full, from 22 to 22¾ ; 4, or Average, from 21½ to 22 ; 3,
or Moderate, from 20¾ to 21½ ; 2, or Small, from 20 to 20¾ ; 1, Below
20. Female heads are half an inch to an inch below these measurements. Those whose heads are —

7, or VERY LARGE. — With quality good, are naturally great ;
with quality and activity 6 or 7, and the intellectual organs 6 or 7, are a
natural genius, a mental giant ; even without education, will surmount
all disadvantages, learn with wonderful facility, sway mind, and become preëminent ; with the organs of practical intellect and the propelling powers 6 or 7, will possess natural abilities of the first order;
manifest a clearness and force of intellect which will astonish mankind, and a power of feeling which will carry all before them ; and,
with proper cultivation, become bright stars in the firmament of intel-

·lectual greatness, upon which coming ages will gaze with delight and astonishment. With quality and activity 5 or 4, are great on great occasions, and, when thoroughly roused, manifest splendid talents, and naturally take the lead among men, otherwise not ; with activity or quality deficient, must cultivate much in order to become much.

LARGE. — With activity and quality 6 or 7, combine great *power* of mind with great activity, exercise a commanding influence over other minds to sway and persuade, and enjoy and suffer in the extreme ; with perceptives 6, can conduct a large business or undertaking successfully, rise to eminence, if not preëminence, and evince great originality and power of intellect, strong native sense, superior judgment, great force of character and feeling, and make a conspicuous and enduring mark on the intellectual or business world, or in whatever direction those superior capacities are put forth. With activity and quality 5, are endowed with superior natural talents, yet require strong incentives to call them out ; undeveloped by circumstances, may pass through life without accomplishing much, or attracting notice, or evincing more than ordinary parts ; but with the perceptive and forcible organs also 6, and talents disciplined and called out, manifest a vigor and energy far above mediocrity ; are adequate to carry forward great undertakings, demanding originality and force of mind and character, yet are rather indolent. With activity only *average*, possess considerable energy of intellect and feeling, yet seldom manifest it, unless brought out by some powerful stimulus, and are rather too indolent to exert, especially *intellect*.

FULL. — With quality or activity 6 or 7, and the organs of practical intellect and of the propelling powers large, or very large, although not really *great* in intellect, or deep, are very clever ; have considerable talent, and that so distributed that it shows to be even more or better than it really is ; are capable of being a good scholar, doing a fine business, and, with advantages and application, of becoming distinguished somewhat, yet inadequate to great undertakings ; cannot sway an extensive influence, nor become really great, yet have excellent natural capacities ; with activity 4 or 5, will do tolerably well, and manifest a common share of talent; with activity only 8, will neither be nor do much worthy of notice. .

AVERAGE. — With activity 6, manifest a quick, clear, sprightly mind, and off-hand talents; and are capable of doing a fair business, especially if the stamina is good ; with activity 7, and the organs of the propelling powers and of practical intellect 6 or 7, are capable o doing a good business, and possess fair talent, yet are not original or profound; are quick of perception; have a good practical understanding; will do well in an ordinary business or sphere, yet never manifest greatness, and out of this sphere are commonplace; with activity only 4, discover only an ordinary amount of intellect; are indisposed and inadequate to any important undertaking; yet, in a common sphere, or one that requires only a mechanical routine of business, can do well; with *moderate or small* activity, will hardly accomplish or enjoy anything worthy of note.

MODERATE. — With quality, activity, and the propelling and perceptive faculties 6 or 7, possess an excellent intellect, yet are more showy than sound; with others to plan and direct, can execute to advantage, yet are unable to do much alone; have a very active mind, and are quick of perception, yet, after all, have a contracted intellect; possess only a fair mental calibre, and lack momentum, both of mind and character; with activity only 4, have but a moderate amount of intellect, and even this too sluggish for action, so as neither to suffer nor enjoy much; with activity 3 or 2, are dull, and hardly *compos mentis*.

SMALL. — Are weak in character and inferior in intellect — indeed, simple or idiotic.

This doctrine, that "size is a measure of power," is equally true of different *groups* of organs, and *regions* of the brain. Those who have a large forehead, with a deficient back and side-head, if of good temperament, will be deep, original *thinkers*, but lack force and energy of character; while those who have heavy base and back-head, with a smaller forehead, will possess energy, courage, passion, sociability. and vim, but lack intellectual capacity. But this point will be eliminated hereafter.

VII. PHRENOLOGY WAS DISCOVERED AND ESTABLISHED BY INDUCTION.

1. This is proved by the entire history of this science as a whole, and of each particular organ and faculty. No part of it rests on theory. In all its parts and details it is wholly a MATTER-OF-FACT science. And any one, by learning the locations and different forms of one or more of its organs, together with their phrenological functions, can test its truth — ascertain for himself whether those noted for special mental gifts or proclivities have or have not the corresponding phrenological developments.

2. All men and animals, as compared with one another, prove that Phrenology expresses a natural ordinance and fact. Man and animals are fashioned upon the same general principles, analogous functions in each being performed by similar organs. Thus all men and all animals see by means of eyes and light, resupply nutrition by means of one common organism, the digestive, all move by muscles, etc. Therefore, if Phrenology is true of any, it must of course be true of all. And their respective Phrenologies, contrasted with one another, and taken in connection with their respective instincts, must needs show whether all were or were not constructed upon phrenological principles. What, then, are the facts?

Phrenology locates the animal propensities at the SIDES of the head, between and around the ears; the social affections in its BACK and lower portion; the aspiring faculties in its CROWN; the moral on its TOP, and the intellectual in the FOREHEAD; the perceptives, which relate us to matter, OVER THE EYES; and the reflectives, in the UPPER part of the forehead. (See cut No. 102.)

Now, since brutes possess at least only weak moral and reflective faculties, they should, if Phrenology were true, have little top-head, and thus it is. Not one of all the following drawings of animals have much brain in either the reflective or moral region. Almost all their mentality consists of the ANIMAL PROPENSITIES, and nearly all their brain is found BETWEEN and AROUND THEIR EARS, just where, according to Phrenology, it should be. Yet the skulls of all human beings rise high above the eyes and ears, and are long on top, that is, have full intellectual and moral ORGANS, as we know they possess these

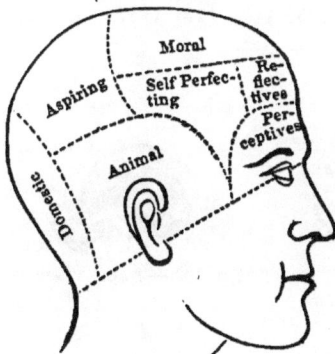

No 102. — GROUPING OF ORGANS.   No. 103. — HUMAN SKULL.

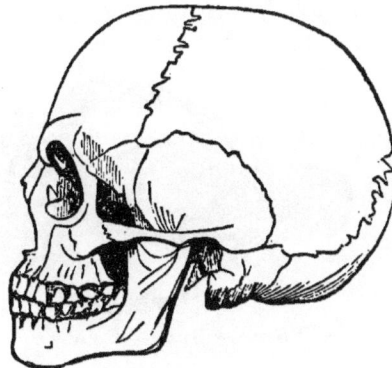

mental ELEMENTS. Compare the accompanying human skull with those of brutes. Those of snakes, frogs, turtles, alligators, etc., slope straight back from the nose; that is, have almost no moral or intellec-

No. 104. — SNAKE.

No. 105. — TURTLE.

tual organs; tigers, dogs, lions, etc., have a little more, yet how insignificant compared with man, while monkeys are between them in both these organs and their faculties. Here, then, is INDUCTIVE proof of Phrenology as extensive as the whole brute creation on the one hand, contrasted with the entire human family on the other.

Again, Destructiveness is located by Phrenology over the ears, so as to render the head wide in proportion as this organ is developed. Ac-

DESTRUCTIVENESS LARGE.

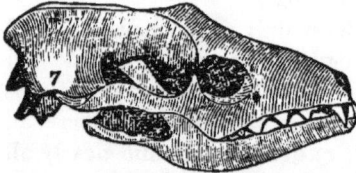

No. 106. — HYENA — SIDE VIEW.

No. 107. — HYENA — BACK VIEW.

cordingly, all carnivorous animals should be wide-headed at the ears, all herbivorous, narrow. And thus they are, as seen in tigers, hyenas, bears, cats, foxes, ichneumons, etc., compared with rabbits, sheep, etc. Contrast cuts 104, 105, 106, 107, 108, 109, 112, 113, 114, 115, 116, 117, 118, and 119, with 110, 111, and 120.

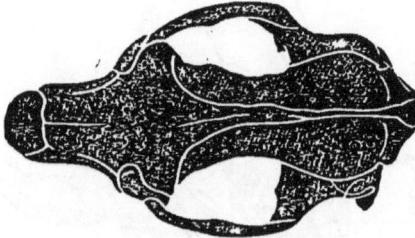

No. 108. — BEAR — TOP VIEW.

No. 109. — BACK VIEW

DESTRUCTIVENESS SMALL.

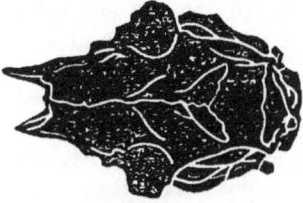 

No. 110.— Sheep — top view.   No. 111.— Rabbit — side view

To large Destructiveness, cats, foxes, ichneumons, etc., add large Secretiveness, both in character and head.

SECRETIVENESS AND DESTRUCTIVENESS BOTH LARGE.

No. 112. — Fox — side view.   No. 113. — Ichneumon — side view.   No. 114.— Do.— back view.

No. 115. — Cat — back view.

   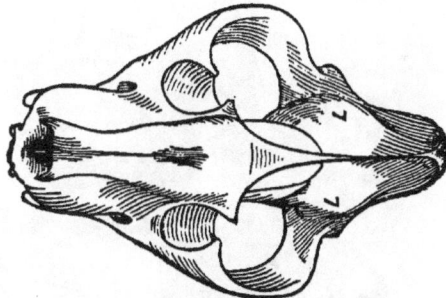

No. 116. — Cat — side view.

No. 117. — Tiger — top view.

Fowls correspond perfectly in head and character with phrenological requisitions. Thus, owls, hawks, eagles, etc., have very wide heads

No. 118. — Owl.   No. 119.— Hawk.   No. 120. — Hen.   No. 121. — Crow

and ferocious dispositions ; while hens, turkeys, etc., have narrow heads, and little Destructiveness in character.   (Cuts 118, 119, 120, and 121.)

The crow (cut 121) has very large Secretiveness and Cautiousness in the head, as it is known to have in character.

Monkeys, too, bear additional testimony to the truth of phrenological science.   They possess in character, strong perceptive powers, but weak reflectives, powerful propensities, and feeble moral elements.   Accordingly, they are full over the eyes, but slope straight back at the reasoning and moral organs, while the propensities engross most of their brain.

The ORANG-OUTANG has more forehead — larger intellectual organs, both perceptive and reflective — than any other animal, with some of the moral sentiments, and accordingly is called the " half reasoning man," its phrenology corresponding perfectly with its character.

No. 122. — INTELLI-
GENT MONKEY.

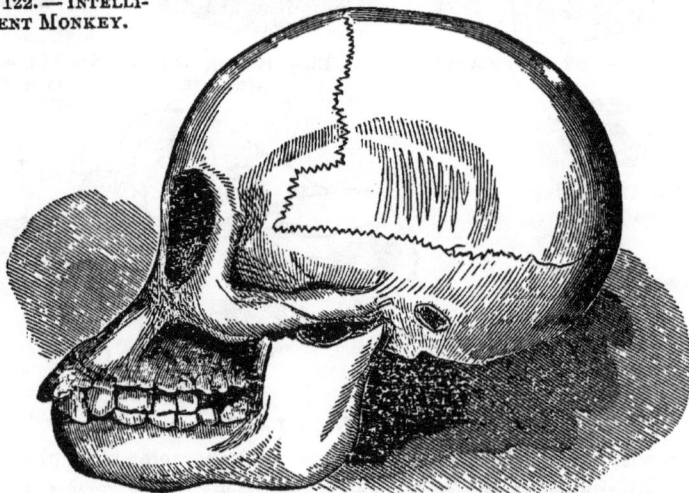

No. 123 — ORANG-OUTANG.
PERCEPTIVES LARGER THAN REFLECTIVES.

The various races also accord with phrenological science.   Thus Africans generally have full perceptives, and large Tune and Language, but retiring Causality, and accordingly possess less reasoning capacity, yet have excellent memories and lingual and musical powers

Indians possess extraordinary strength of propensities and perceptives but moderate moral or inventive power ; and, hence, have very wide round, conical and rather low heads, but are large over the eyes.

Indian skulls can always be selected.from Caucasian, just by these developments; while the Caucasian race is superior in reasoning power and moral elevation to all the other races, and accordingly, has a higher and bolder forehead, and a more elevated and elongated top head.

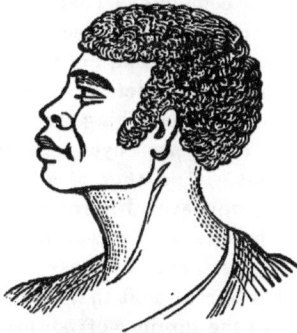

No. 124. — AFRICAN.

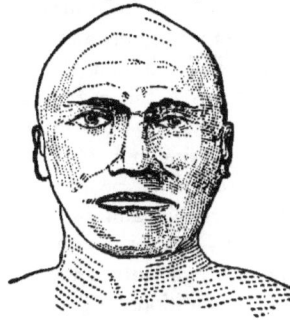

No. 125. — INDIAN CHIEF.

Finally, contrast the massive foreheads of all giant-minded men — Bacons, Franklins, Miltons, etc., with the low, retiring foreheads of idiots. In short, every human, every brutal head, is constructed

LARGE AND SMALL INTELLECTUAL REGION.

No. 126. — BACON.

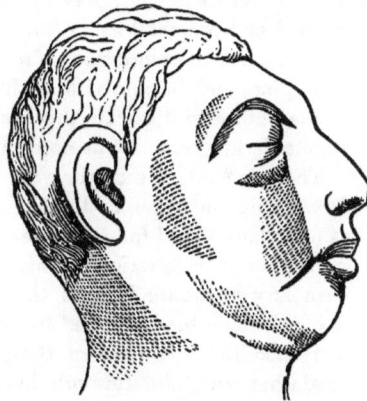

No. 127. — IDIOT.

throughout strictly on phrenological principles. Ransack air, earth, and water, and not one palpable exception ever has been, ever can be, adduced. This WHOLESALE view of this science precludes the possibility of mistake. Phrenology is therefore a PART AND PARCEL OF NATURE — A UNIVERSAL FACT.

VIII. The states of all organs and functions are in reciprocal rapport.

In the very nature and fitness of things the correspondence between
all organs and their functions must be and is complete. That is, the
states of all organs and of their respective functions must be reciprocal.
What means it that the stomach is the organ of digestion, but that all
the states of this organ correspond with those of its functions ? How
could the eye be the organ of vision unless all the changing states of
this eye similarly affect the sight? How could poor eyes execute
good functions, or good eyes poor functions ? And thus of all the
other organic and functional states. Thus, whenever Nature would
put forth *power* of function, she does so by means of power in the
organ which puts it forth. And so of quickness, and all other functional conditions. Thus the office of wood is to rear aloft that stupendous tree-top, and hold it there in spite of all the surgings of powerful
winds upon its vast canvas of trunk, limbs, leaves, and fruit. Now
this requires an immense amount of power, especially considering the
great mechanical disadvantage involved. This power Nature supplies,
not by bulk, because this, by consuming her material and space, would
prevent her making many trees, whereas her entire policy is to form
all the trees she can ; but by rendering the organic *texture* of wood
as solid and powerful as its function is potential. And the more solid
its structure, the more powerful its function, as seen in comparing oak
with pine, and lignum vitæ with poplar. But, letting this single example suffice to illustrate this law, which obtains throughout the entire
vegetable kingdom, let us apply it to the animal.

The elephant, one of the very strongest of beasts, is so powerful in
dermis, muscle, bone, and entire structure, that bullet after bullet shot
at him, flatten, and fall, harmless at his feet. The lion, too, is as strong
in texture as in function. Only those who know from observation can
form any adequate idea of the wiry toughness of those muscles and
tendons which bind his head to his body, or of the solidity of his bones ;
corresponding with the fact that, seizing a bullock in his monster jaw,
he dashes with him through jungle and over ravine, as a cat would
handle a squirrel. And when he roars, a city trembles. The structures of the white and grizzly bear, of the tiger, hyena, and all powerful animals, and, indeed, of all weak ones, in like manner correspond
equally with their functions. All quickness of function is put forth
by quick-acting organs, all slowness by the slow ; and thus of all or

gans and functions throughout every phase and department of universal life and nature. Indeed, in and by the very nature of things this correspondence *must* exist. For how *could* weak organs possibly put forth powerful functions, or slow organs quick functions? In short, this correspondence between organic conditions and functions is fixed and absolute — is necessary, not incidental, — is universal, not partial, — is a relation of cause and effect, and governs every organ and function throughout universal life and nature.

Governs you and I, reader. And in all our functions. If, in the plenitude of Divine Wisdom, man had been created a purely mental being, he would have needed no body, and could not have used one; whereas, instead, he has been created a compound being, composed of both body and mind. Nor are those seemingly opposite entities strangers to each other. Instead, they are inter-related by ties of the most perfect reciprocity — so perfect that every conceivable condition of either reciprocally affects the other. How can weak muscles put forth strength, or a sluggish brain manifest mental activity? Hence, to become great, one must first become *strong* — and in the special *organs* in whose functions he would excel. Would you become great mentally, then first become strong cerebrally. Or, would you render that darling boy a great man, first make him a *powerful animal*. Not that all powerful animals are great men, but that all great men are, and must needs be, powerful animals. Our animal nature is the basis of all our mental and moral functions. It so is in the very constitution of things, that mind can be put forth *only* in and by its material organism, and is strong or weak, quick or sluggish, as its organism is either.

HEREDITARY ORGANISM AS AFFECTING MENTALITY.

Hereditary organic quality is the first, basilar, and all-potent condition of all power of function, all happiness, all everything. This is congenital — is imparted by the parentage along with life itself, of which it is the paramount condition and instrumentality. It depends mainly on the original nature of the parents, yet partly also on their existing states of body, mind, and health, their mutual love or want of it, and on other like *primal* life-conditions and causes. It lies behind and below, and is infinitely more potential than education and all associations and surrounding circumstances, — is, in short, what renders the grain cereal, the oak oaken, fish fishy, fox foxy, swine swinish, tiger tigerish, and man human. See this whole subject fully discussed in the author's new work entitled " The Family."

Each creature much resembles a galvanic battery, and its life-force depends mainly on how that battery works. And this on those congenital conditions which *establish* life — a subject infinitely important, and generally overlooked, but treated fully in " Supplement to ' the family ' " or " Offspring and their Hereditary Endowment."

These organic conditions cannot well be described, hardly engraved, but are easily perceived by a practiced eye. They are quite analogous to temperament, on which little has yet been written, but lie behind and below all temperaments — are, indeed, their determining cause. Some of their signs are coarseness and fineness of hair, skin, color, form, notion, general tone of action and mental operation, etc. A comparison of the following engravings of Fanny Forester with the idiot

Emerson will give some outline idea of this point. A still better is found in comparing man with animal. In fact, the main differences between vegetables and animals, as compared among one another, and all as compared with man, and different men as compared with each other, as well as the entire style and cast of character and sentiment, everything, is consequent on these organic conditions — in short, is what we call " bottom " in the horse, " the

No. 128.—FANNY FORESTER.

breed " in full-blooded animals, and " blood " in those high and nobly born. Those marked 1

7. —. Are preëminently fine-grained, pure-minded, ethereal, sentimental, refined, high-toned, intense in emotion, full of human nature, most exquisitely susceptible to impressions of all kinds, most poetic in temperament, lofty in aspiration, and endowed with wonderful intuition as to truth, what is right, best, etc.; are unusually developed in the interior, or spirit-life, and far above most of those with whom they come in contact, and hence find few congenial spirits, and are neither understood nor appreciated ; when sick, suffer inexpressibly, and if children, are precocious — too smart, too good to live, and absolutely must be treated physiologically, or die early.

1 Hereafter, the words ' those marked " will be omitted, and the description begin " 7 - - Are," etc.

6. — Are like 7, only less so; are finely organized, delicate, suscep-tible, emotional, pure-minded, intellectual, particular, and aspiring after a high state of excellence; full of human nature, and true to its intuitions and instincts; have a decided predominance of the mental over the physical; are able and inclined to lead excellent human lives, and capable of manifesting a high order of the human virtues.

5. — Are more pre-inclined to the good than bad, to ascend than de-scend in the human scale; can, by culture, make excellent men and wom⌀n, but require it; and should avoid those habits which clog or deprave the mental manifestations, and, to attain superiority, must " strive for it."

4. — Are simply fair in organic tone; are good under good sur-roundings, but can be misled; must avoid all deteriorating habits and causes, spirits and tobacco, bad associates, etc.; assiduously cultivate the pure and good, and study to discipline intellect, as well as purify the passions, and rely the more on culture and a right physi-ological life, because the hereditary endowment is simply fair.

No. 129.—EMERSON, AN IDIOT.

3. — Are rather lacking in or-ganic quality, and better adapted to labor than study; rather sluggish mentally, and given to this world's pleasures; had but a commonplace parentage; need to be strictly temperate in all things, and avoid all forms of temptation, vulgar associates in particular, and make up by the more assiduous cultivation what has been withheld by nature.

2. — Are coarse grained in structure and sentiment, and both vulgar and non-intellectual; had poor parental conditions; are low, grovel-ing, and carnal, as well as obtuse in feeling and intellect; are poorly organized, and incapable of high attainments; hence restrain the pas-sions, and cultivate intellect and the virtues as much as possible, and especially avoid alcoholic liquors, tobacco, and low associates.

1. — Are really foolish, and *non compos mentis.*

To CULTIVATE. — First, guard against all perversion of the facul-ties, all forms of intemperance, tobacco, over-eating, pork, rich pastry, especially late suppers ; be much of the time in the open air ; work and exercise abundantly; bathe daily, and keep the body in just as good condition as possible; mingle with the high and good; exercise all the faculties assiduously, in the best possible manner, and in strict accordance with their natural functions; cultivate a love of nature, art, beauties, and perfections — in short, encourage the good, true, and right, and avoid the bad.

To RESTRAIN. — Cultivate a love of the terrestrial — of this world, its pleasures and luxuries, — for you require animalizing. You live too much in the ideal. Live more with the actual and tangible. Callous yourself against much that now abrades your finer sentiments, and shrink not from contact with those not quite up to your standard. You are adapted to a more advanced state of humanity, but should come down to the present and material. Above all, do not be too fastidious, qualmish, or whimmy, but make the best of what is ; cling to life, and enamor yourself of its objects and pleasures.

Closely related to these organic conditions is —

HEALTH — ITS VALUE, CONDITIONS, AND RESTORATION.

Health consists in the normal and vigorous exercise of all the physical functions, and disease in their abnormal action. Health is pleasurable, disease painful. Health is life, for life consists in the normal action of those same functions in which health consists. And to improve health is to increase life itself, and all its pleasures. Some writer has appropriately defined health thus : —

Planting your foot upon the green sward, looking around, and yielding yourself to whatever feelings naturally arise, health is proportionate to that buoyant, jubilant, exhilarating, ecstatic feeling which supervenes. It is to all our functions what motive power is to machinery — sets them off with a rush and a bound. It both makes us happy, and causes everything else to increase that happiness.

But disease renders us miserable, and turns everything around us into occasion of misery. It both weakens and perverts our mental being. Indeed, health is the quintessence of every earthly good — disease of every terrestrial evil. Poor indeed is he, however rich in money, in honors, in office, in everything else whatsoever, whose *health* is poor ; for how can he enjoy his dollars and honors ? But rich indeed is he who is healthy, however poor in money, for he *enjoys* whatsoever he has or is. A rich man may, indeed, purchase a luxuriant dinner, but without health does not, cannot *relish* it ; whereas a poor man, with health, enjoys even a dry crust.

The rich need health to enjoy their riches; the poor doubly, in order to prevent becoming poorer. But to be poor *and* sickly is the uttermost of human evil. Nor can the poor afford to be sick; for their health is their *all*, to themselves and families. Nor should they allow *any* thing whatsoever to impair it, but make *health paramount*.

Even the very talents of men depend mainly on health. Is not the brain confessedly the organ of the mind? Now, what means it, that the brain is the organ of the mind, but that all *its* conditions similarly affect the mentality? And since all the states of the body and brain act reciprocally — consequent on that vast network of nerves which ramify throughout every part and parcel of the body, and terminate in the brain, — of course all existing conditions of the body similarly affect these nerves, and thereby the brain, and therefore the mind, rendering all the states of either body or mind reciprocal with those of the other. Is the body sick, or weak, or exhausted, or inflamed, or sleepy, or exhilarated, is not the mind equally so? Then to originate great thoughts, or to conceive pure and exalted sentiments, must not the *brain* be in a vigorous state? And in order to acquire cerebral vigor, must not all the bodily functions be equally vigorous? And to this end, must not those health-laws which cause this vigor be observed? Of what avail the learning of the sickly scholar, the talents of the invalid, or the goodness of the pious dyspeptic? They can *do* nothing, can *enjoy* nothing — are but burdens to themselves and friends. Can we think, or remember, or study without that energy furnished by the body? No more than move machinery without motive power. How, then, can that boy become a great or learned man without possessing physical vigor? Or that delicate and beautiful girl a capable or good woman, wife, or mother without possessing animal vigor? Let it be forever and everywhere remembered, that both judgment *and* memory, reason and poetry, eloquence and philosophy, even morality and religion, all the virtues and all the vices — in short, one and all of the human functions, are carried forward by animal power. Even the very sensual pleasures of the debauchee are exercised by this very animal force, and grow weak when and because it declines. And as physical power depends on the observance of certain physical laws, the violation of which weakens both body and mind, of course the first duty of every human being to himself and Creator — of parents to their children, of ministers to people, writer to reader, and one to all — is to —

LEARN AND OBEY THE HEALTH LAWS.

And on this point is just where our whole educational system — collegiate especially — is radically defective. It eclipses more genius by weakening the body than it eliminates by study. Children are always smarter and better relatively than adults, because injured by that false educational system which impairs mind, memory, and morals by breaking down good physical constitutions. The Romans appropriately named their schools " gymnasia," from those muscular *exercises* which both formed their leading feature, and secured a strong mind, by strengthening the body. Our schools and colleges are, and will continue to be, fundamentally defective, till remodeled upon the basis of *health as a means* of scholarship and talents.

Nor intellect merely, but our very *morals* and piety, depend on health. Can we even pray or worship without vitality? And what is more, the very vices of mankind are consequent mainly on the infringement of the *physical* laws.

Hereditary conditions in parents cause depravity in their children; yet even they do it by deranging the body. It is what men eat and drink, it is how they live, sleep, etc., it is their *physiological* conditions and habits, that cause nine-tenths of human depravity. Are not both children and adults depraved when cross, and cross because sick ; that is, rendered sinful by being unwell? Who does not know that drunkenness engenders depravity — makes the best men bad ? But why, and how ? By disordering the body. And since by alcohol, why not by tobacco, gluttony, and every other wrong physical state ? Are not drunkenness and debauchery concomitants ? Are not dyspeptics always irritable ? The truth is, that all abnormal physical action causes abnormal mental action, which is sin. To become good and answer the end of their being, men must *live* right, must learn to eat right, and sleep, exercise, bathe, breathe, etc., in accordance with nature's requisitions. And nine-tenths of the sinfulness of mankind has this purely physical origin, and can be cured by physical means.

Health is the natural state of man, animal, vegetable, all that lives — is the ultimate of life. Like all else in nature, it has its laws; and these laws obeyed, will render it perfect from birth to death. It even requires immense violation of these laws seriously to impair it. Bird and beast are rarely unhealthy, except when rendered sickly by man Has our benevolent Creator granted this greatest of boons to beasts, but denied it to man ? No. None need ever be sick, for there are health laws, which, if obeyed, guarantee the very perfection of health To become sickly is foolish ; for it cuts off every pleasure, and induces every ill — is even wicked, for it is consequent only on a violation of the laws of our being, and all violation of law is sin. And the health laws are as much laws of God — written by his finger on our very constitution — as the Decalogue. In short, none have any right to be sick. It is alike the privilege, as it is the sacred duty of one and all to be and keep well ; that is, to observe the health laws. And of parents to keep their children well.

EXISTING STATES OF HEALTH, AND ITS IMPROVEMENT.

While this condition has a most important influence on both the quantity and quality of all the mental manifestations, yet to mark it correctly, without aid from those examined, is exceedingly difficult. It may seem good, when actually poor, because its functions may be exhilarated by inflammation, which both perverts and weakens ; or it may seem much poorer than it really is, because of merely temporary debil- ·ity, while the heart's core remains sound. But its serious impairment leaves all the functions, phrenological included, proportionally less vigo- rous than the sizes of their organs indicate. Those who have health —

7. — Are full to overflowing with life, buoyancy, light-heartedness, and ecstasy ; are strong and lively ; enjoy food, sleep, action, nature, all the physical functions, to the highest degree; rarely ever have a pain or ache, or become tired ; can do and endure almost any and every- thing ; withstand miasma and disease remarkably; recuperate readily; experience a certain gush, glow, vivacity, and briskness in the action of all the faculties ; as well as the highest and most perfect flow and exercise of each of the life-functions.

6. — Are healthy and happy ; exercise all the organs with vigor and power; turn everything into pleasure, and dash off trouble as if a mere trifle, and yet can endure any amount of pain and exposure ; feel jubi- ,ant and joyous year in and year out ; and do everything easily, all the functions being condensed and hearty, and the whole being full of snap and life.

5. — Have a good, full share of life-force, vigor, and vivacity — of health, happiness, desire and ability to perform, enjoy, and accomplish ; can stand a good deal, but must not go too far, and have sufficient stamina for all practical purposes, but none to spare or waste foolishly.

4. — Have fair, average health, if it is well cared for, yet are some- times subject to ailments ; are in the main healthy and happy, but

must live regularly ; experience rather a tame, mechanical action of all the faculties, instead of that zest and rapture imparted by perfect health ; can accomplish and enjoy much, but must take things leisurely ; if careful, can live and wear on a long while yet, but if careless, are liable to break down suddenly and finally ; and become irritable, dissatisfied, dull, forgetful, and easily fatigued, and must cherish what health remains.

3. — Are deficient in animation and recuperative power, and feel tired and good for nothing most of the time ; with activity 6 or 7 are constantly overdoing, and working up in mental or physical action those energies which ought to go to the restoration of health, not to labor ; need abundance of rest and recreation, and give out at once if deprived of sleep; must stop all unnecessary vital drains, such as chewing, smoking, drinking, late hours, and all forms of dissipation, and should manufacture all the vitality possible, but expend the least.

2. — Are weakly, sickly, and inert; feeble in desire and effort ; capable of enduring and enjoying but little ; live a monotonous, listless, care-for-nothing, half-dead-and-alive life, and must either restore health or give up, and enjoy comparatively nothing.

1. — Have barely life enough to keep soul and body together ; are just alive, and have almost lost life's pleasures, powers, desires, and aspirations.

To Cultivate. — First ascertain what causes your disease or debility ; if heart, lungs, muscles, stomach, etc., are marked low, apply special culture to the weak organs — see the cultivation of each, — and assiduously study the health laws, and conscientiously fulfill them, making everything else subservient thereto. Especially take extra pains to *supply vitality*, but waste none in any form of excess.[1]

Restrain You Need Not. — Health cannot be too good. When, however, you find a surplus of animal vigor, work it up in one or another of life's ends and efforts.

The Temperaments.

This term has long been employed to designate certain physical constitutions as indicative of certain mental characteristics. The idea expressed in our definition of " hereditary organism " is quite like that of the temperaments. They were formerly classified thus : The ner-

[1] For a complete and detailed view of health-culture, see the Author's new work, entitled — " Health : its Value, Natural Laws, Conditions, Preservation and Restoration; including the Organism, the Temperaments " etc.

vous, indicated by light complexion, large brain, and smaller stature, and indicating superior talents, refinement, and scholarship ; the bilious, indicated by dark complexion, large bones, powerful muscles, prominent features, and a large and spare form, and indicating a supposed surplus of bile, irritability, violence of passion, and melancholy, along with strength of character ; the sanguine, indicated by a florid complexion, sandy hair, blue eyes, fullness of person, and abundance of blood, and indicating warmth, ardor, impulsiveness, and liability to passional excesses ; and the lymphatic, indicated by full, plethoric habit, distended abdomen, excessive adipose deposit, and indicating a good, cosy, lax, enjoying disposition, with a stronger proclivity to sensuous pleasures, rather than intellect or action of any kind. But this classification is practically discarded, without its place having been supplied. The doctrine of the temperaments in full remains unwritten. Meanwhile we propound the following

CLASSIFICATION AND DEFINITIONS.

Man is composed physically of three great classes of organs, the predominance or deficiency of each of which is called a predominant or deficient temperament, each giving a particular form to the body — shape being its index, — and likewise a particular set of phrenological developments, and consequent traits of character. That is, given forms of body indicate and accompany special talents, dispositions, and mental proclivities ; and the art in delineating phrenological character depends in a great degree on reading correctly the temperaments and organic conditions, and their controlling influences on character ; for they exert, as it were, the ground-swell as to the direction and action of the phrenological manifestations. Thus Causality, with the vital temperament-predominant, takes on the phase of planning, of common sense, of reasoning on matter, of adapting ways and means to ends, etc. But with the nervous or mental predominant, the same sized Causality manifests itself in logic, metaphysics, investigation, the origination of ideas, in intellectual clearness and power, etc. And it requires the sharpest eye and clearest head in the examiner to discern the bearings and influences of these temperamental and organic conditions on the intellectual and moral manifestations. And the mistakes of amateurs, of connoisseurs even, are more temperamental than phrenological. Still they are sometimes consequent on health conditions. Thus the same person in one state of health is irritable, violent, passional, perhaps even sensual and wicked, who in another physical condition is amiable, even-tempered, moral, and good. A given amount of ideality is much more ideal, of language much more expressive, of the affections more affectional, and moral tone more lofty, in combination with the mental temperament than vital. But our proposed limits do not allow us to extend our observations. Still, the following descriptions give the outline, and put inquirers on the track of further observations.

THE VITAL TEMPERAMENT.

This embraces the heart, lungs, stomach, liver, bowels, and that entire system of internal organs which creates life-force. It is **very** large in William G. Hall.

No. 130. — WILLIAM G. HALL.

The large end of a good egg is warmer than its other parts, because its vitality resides there ; but, this cold, life is extinct. Incubate it a short time, and break the shell at this end, and you will find the heart palpitating and blood-vessels formed — the yolk furnishing the required nutrition. The vital apparatus forms first, and deposits the material for forming the other portions ; is more active during juvenility than the other parts ; sustains the whole animal economy ; is the source of all power and energy ; creates animal heat ; resists cold and heat, disease and death ; and resupplies muscle, brain, and nerve with that life power expended by their every exertion. It is to the man what fire, fuel, water, and steam are to machinery — the *vis animœ*, the *primum mobile* — the first great prerequisite of life itself and all its functions.

Its decided predominance is accompanied by a round head, well developed at the base, large Amativeness, Acquisitiveness, Alimentiveness, Benevolence, and Language ; large organs of the animal propensities generally ; a rapid widening of the head from the corners of the eyes to the tips of the ears ; side-head spherical and well filled out ; forehead generally full or square, and broad rather than high ; perceptive organs large, and all the organs short and broad rather than long or pointed.

7. — Are fleshy ; short and broad built ; stocky ; deep and large chested ; broad and round shouldered ; impetuous ; impulsive ; enthusiastic ; hearty ; good livers ; fond of meats, condiments, stimulants, and animal pleasures ; have a strong, steady pulse ; large lungs and nostrils ; a full habit ; florid complexion ; flushed face ; light or sandy hair or whiskers ; sound and well-set teeth ; great endurance of fatigue, privation, and exposure ; great love of fresh air, out-of-door exercise, and physical action, but not of *hard* work ; a restlessness which can not endure in-door confinement, but must be abroad, and constantly doing something ; great zeal, ardor of desire, and more practical common sense than book-learning ; more general knowledge of men and things than accurate scientific attainment ; more shrewdness and off-hand talent than depth ; more availability than profundity ; and love of pleasure than power of thought.

6. — Are like 7, though not in as great extremes ; generally fleshy and of good size and height, if not large ; well-proportioned ; broad-shouldered ; muscular ; prominent and strongly-marked in features ; coarse and homely ; stern and harsh ; strong, but often awkward, and seldom polished ; best adapted to some laborious occupation, and enjoy hard work more than books or literary pursuits ; have great power of feeling, and thus require much self-government ; possess more talent than they can exhibit to others ; manifest mind more in business, in creating resources and managing matters, than in literary pursuits, or mind as such ; prefer some light, stirring, active business, but dislike drudgery ; turn everything, especially bargains, to good account ; look out for self ; get a full share of what is to be had ; feel and act out.

"every man for himself," and are selfish enough, yet abound in good feeling; incline to become agents, overseers, captains, hotel keepers, butchers, traders, speculators, politicians, public officers, aldermen, contractors, etc., rather than anything requiring steady or hard work· and are usually healthy, yet very sick when attacked, brought at once to the crisis, and predisposed to gout, fevers, apoplexy, congestion of the brain, etc.

5. — Have a good share of life-force, yet none to spare; withstand a good deal, yet must not waste vitality, and should live in a way to improve it.

4. — Have sufficient vitality to sustain life, and impart a fair share of energy to the functions, but by no means sufficient to put forth their full power, and should make its culture a first life-object.

3. — Are rather weakly and feeble ; often half prostrated by a feeling of languor and lassitude; can keep doing about all the time if slow, and careful not to overdo, the liability to which is great when Activity is 6 or 7; need much rest; cannot half work, or enjoy either body or mind; suffer much from fatigue and exhaustion, and would be glad to do, but hardly feel able.

2. — Are too weak and low to be able either to do, enjoy, or accomplish much; should both give the vital organs every possible facility for action, and also husband every item of vitality; be extremely careful not to overwork, and spend much time in listless, luxuriating ease, while nature restores the wanting vitality.

1. — Are almost dead from sheer inanition.

To Cultivate. — Ascertain which of the vital organs is deficient, and take all possible pains to improve its action; see directions for increasing the action of the heart, lungs, stomach, etc.; alternate with rest and exercise; "away with melancholy," banish sadness, trouble, and all gloomy associations, and cultivate buoyancy and light-heartedness; enjoy the present, and make life a glorious holiday instead of a weary drudgery; if engaged in any confining business, break up this monotony by taking a long leave of absence — a trip to Lake Superior, California, or Europe, a long journey, by horticulture, or parties, or frolicking with children; by going into young and lively society, and exercising the affections; bringing about as great a change as possible in all your habits and associations. Especially cultivate a love of everything beautiful and lovely in nature, as well as study her philosophies ; bear patiently what you must, but enjoy all you can ; keep doing all you are able, but other things than formerly, and what interests you. You should watch and follow your intuitions or instincts and if you feel a special craving for any kind of food or pleasure, indulge it. Especially be regular in sleep, exercise, eating, and all the vital functions, as well as be temperate in all things. Above all keep your mind toned up to sustain the body. Aid your weak organs by will-power, that is, bring a strong will to aid digestion, breathing, etc., and keep yourself up thereby. Determine that you *won't* give up to weakness or death, but will live on and keep doing in *spite* of debility and disease. Fight life's battles like a true hero, and keep the head cool by temperance; the feet warm by exercise; the pores and evacuations open by ablution and laxative food; and heart warm by cherishing a love of life and its pleasures. And don't fail to keep up a gentle pounding and frequent brisk rubbing of chest, abdomen, and feet, so as to start the mechanical action of the visceral organs. Nothing equals this for revivifying dormant or exhausted vitality, and none are too poor or too much occupied to avail themselves of it.

To RESTRAIN. — Those who manufacture vitality faster than they expend it, are large in the abdomen; too corpulent; even obese; often oppressed for breath; surcharged with organic material; too sluggish to expend vitality as fast as it accumulates, and hence should work, work, work, early and late, and with all their might, and as much as possible with their muscles and out-of-doors; should eat sparingly, and of simple food; avoid rich gravies, butter, sweets, fat, and pastry, but live much on fruits; sleep little; keep all the excretory organs free and open by an aperient diet, and especially the skin by frequent ablutions, the hot bath, etc.; breathe abundantly, so as to burn up the surplus carbon; sit little, but walk much; never yield to indolence; work up energy by hands and head, business and pleasure, any way, every way, but keep consuming vitality as fast as possible. Some fleshy persons, especially females, give up to indolence and inanity; get "the blues," and lounge on rocking-chair and bed. What is wanted is to *do*, not to loiter around. *Inertia* is your bane, and action your cure. If flushed, feverish, nervous, etc., be careful not to overdo, and rely on *air*, warm bath, and gentle but continued exercise, active or passive, but not on medicines.

THE LUNGS — BREATHING.

All that lives, down even to vegetables and trees, breathes; must breathe in order to live; live in proportion as they breathe; begin life's first function with breathing, and end its last with their last

breath. And breathing is the most important function both from
first to last, because the grand stimulator and sustainer of all. Would
you get and keep warm when cold, breathe copiously, for this increases
that carbonic consumption all through the system which creates all
animal warmth. Would you cool off and keep cool in hot weather,
deep, copious breathing will burst open all those myriads of pores,
each of which, by converting the water in the system into perspira-
tion, casts out heat, and refreshes mind and body. Would you labor
long and hard, with intellect or muscle, without exhaustion or injury,
breathe abundantly ; for breath is the great reinvigorator of life and
all its functions. Would you keep well, breath is your great preven-
tive of fevers, of consumption, of " all the ills that flesh is heir to."
Would you break up fevers, or colds, or unload the system of morbid
matter, or save both your constitution and doctor's fee, cover up
warm, drink soft water — cold, if you have a robust constitution, sufficient to produce a reaction ; if not, use hot water — then breathe,
breathe, breathe, just as fast and as much as possible of fresh air, and
in a few hours you can " forestall and prevent " the worst attack of
disease you ever can have; for this will both unload disease at every
pore of skin and lungs, and infuse into the system that *vis animæ*
which will both grapple with and expel disease in all its forms, and
restore health, strength, and life. Nature has no panacea like it. Try
the experiment, and it will revolutionize your condition. And the
longer you try, the more it will regenerate your body and your mind.
Even if you have the blues, deep breathing will soon dispel them,
especially if you add vigorous exercise. Would you even put forth
your greatest mental exertions in speaking or writing, keep your
lungs clear up to their fullest, liveliest action. Would you even
breathe forth your highest, holiest orisons of thanksgiving and worship
deepening your inspiration of fresh air will likewise deepen and
quicken your divine inspiration. Nor can even bodily pleasures be
fully enjoyed except in and by copious breathing. In short, deep,
copious breathing is the alpha and omega of all physical, and thereby
of all mental and moral function and enjoyment.

7 and 6. — Have either a full, broad, round chest, or a deep one, or
both ; breathe freely, but rather slowly; fill the lungs clear up full at
every inspiration, and empty them well out at every expiration; are
warm, even to the extremities; red-faced; elastic; buoyant; rarely
ever subject to colds, and cast them off readily; feel buoyant and ani-
mated, and are thus capable of great vigor in all the functions,
physical and mental.

5 and 4. — Are neither pale nor flushed, neither ardent nor cold, but a little above medium in these respects, and somewhat liable to colds.

3. — Breathe little, and mainly with the top of the lungs ; move the chest but little in breathing, and the abdomen less, perhaps none at all; are often pale, yet sometimes flushed because feverish; frequently do and should draw in long breaths; are quite liable to colds and coughs, which should be broken up at once, or they may induce consumption; often have blue veins and goose-flesh, and are frequently tired, listless, and sleepy, and should take particular pains to increase lung action.

2. — Are strongly predisposed to lung diseases; have blue veins and sallow complexion, and are very subject to coughs and colds; are often dull, and always tired; frequently catch a long breath, which should be encouraged by making all the breaths long and frequent; are predisposed to consumptive diseases, but can stave them off, provided proper means are adopted: break up colds as soon as they appear, and take particularly good care of health.

1. — Have barely lung action enough to live, and every function of body or mind is poorly performed.

To Cultivate. — First and mainly breathe deeply and rapidly ; that is, draw long and full breaths; fill your lungs clear up full at every inspiration, and empty them out completely at every expiration ; not only heave the chest in breathing, but work the abdomen. To do this, dress loosely and sit erect, so that the diaphragm can have full play ; begin and keep up any extra exertion with extra lung action; often try how many deep and full breaths you can take ; ventilate your rooms, especially sleeping apartments, well, and be much of the time in the open air; take walks in brisk weather, with special reference to copious respiration; and everywhere try to cultivate full and frequent lung inflation, by breathing clear out, clear in, and low down ; that is, make all your breathing as when taking a long breath.

THE CIRCULATION.

" For the blood thereof is the life thereof." The blood is the great porter of the system ; carries all the material with which to build up and repair every part, and hurries off all the waste material, which it expels through lungs and skin.

And the heart is one circulatory instrumentality. Without heart, even lungs would be of no account, nor heart without lungs. They are twin brothers, are co-workers at the very fountain-head of life and all its energies. Even diseased organs are unloaded of morbid matter, reanimated, and rebuilt mainly by blood. Blood good or poor, the whole system, brain and mind included, is in a good or poor condition; but blood wanting, all is wanting; heart poor, all is poor; heart improved, all is improved.

7 and 6. — Have an excellent and uniform circulation, and warm hands, feet, and skin ; never feel chilly ; withstand cold and heat well ; perspire freely ; have a slow, strong, steady pulse, and are not liable to sickness.

5 and 4. — Have a fair, yet not remarkably good, circulation, and generally, though not always. warm hands and feet ; are not much pinched by cold; perspire tolerably freely, yet better if more ; and need to promote circulation, at least not impede it.

3. — Have but poor circulation, along with uneasiness and palpitation of the heart ; are subject to cold hands and feet, headache, and a dry or clammy skin ; find the heart to beat quicker and stronger when drawing than expiring breath ; are chilled by cold, and overcome by hot, weather ; are subject to palpitation of the heart on any extra exertion, walking fast or up stairs, or a sudden startle, etc., and very much need to *equalize* and *promote* the circulation.

2. — Have weak circulatory functions, and either a fluttering pulse, very fast and very irregular, or it is weak and feeble ; suffer from chilliness, even in summer ; are very much affected by changes in the weather; very cold in the extremities, and suffer much from headache, and heat and pressure on the brain ; are subject to brain fever, and often a wild, incoherent action of the brain, because the blood which should go to the extremities is confined mainly to the head and vital organs; feel a sudden pain in the head when startled or beginning to put forth any special exertion, and suffer very much mentally and physically from heart affections and their consequences.

1. — Have scarcely any pulse, and that little is on a flutter ; are cold, and " more dead than alive."

To CULTIVATE. — Immerse hands and feet semi-weekly in water as hot as can be borne, ten minutes, then dash on or dip into cold water, and rub briskly, and heat by the fire till warm. and follow with active exercise, breathing at the same time according to directions just given ; if there is heat or pain about the heart, lay on a cloth, wrung out of cold water at night ; rub and pat or strike the chest on its upper and left side, and restrain appetite if it is craving, and cultivate calmness and quiet. If sufficient vitality remains to secure reaction, putting the feet in ice-cold water will be of great service.

To RESTRAIN is not necessary, except when excessive circulation is consequent on disease, in which case remove the cause. A healthy circulation cannot be too great.

ALIMENTATION.

By that truly wonderful process, digestion, food and drink are made to subserve intellect and moral sentiment — converted into thought and emotion. Then, must not different kinds of food produce different mental and moral traits? A vast variety of facts answer affirmatively. Rollin says that pugilists, while training for the bloody arena, were fed exclusively on raw meat. Does not the food of lion, tiger, shark, eagle, etc., re-increase their ferocity, and that of deer, dove, and sheep re-double their docility? Does not this principle explain the ferocity of the Indian, force of the Anglo-Saxon, and subserviency of the Hindoo? Since alcoholic drinks excite the animal passions more than the intellectual and moral faculties, why not also meat, condiments, and all stimulating food as well? And why not vegetables and the cereals, by keeping the system cool, promote mental quiet, intellectual clearness, and moral elevation? At all events, less meats and more vegetables, grains, and fruits would render men less sensual, and more talented and good. And those who would become either, must mind what and how they eat.

STOMACH. — 7. — Can eat anything with impunity, and digest it perfectly; can live on little, or eat much, and need not be very particular as to diet.

6. — Have excellent digestion; both relish and dispose of food to perfection; are not liable to dyspepsia; have good blood and plenty of it, and a natural hearty appetite, but prefer the substantials to knick-nacks; hate a scanty meal, and have plenty of energy and good flesh.

5. — Have good, but not first-rate digestion, and it will continue good till bad eating impairs it, still must not invite dyspepsia, by bad living.

4. — Have only fair digestive vigor — too little to be abused — and need to promote it.

3. — Have a weak digestive apparatus, and variable appetite — very good, or else very poor; are a good deal pre-inclined to dyspepsia often feel a goneness and sinking at the stomach, and a general lassitude and inertia; sleep poorly, and feel tired and qualmish in the

morning ; have either a longing, hankering, pining, hungry feeling, or a loathing, dainty, dormant appetite ; are displeased and dissatisfied with everything ; irritable and peevish, dispirited, discouraged, gloomy, and miserable ; feel as if forsaken and neglected · are easily agitated, and oppressed with an indefinable sense of dread, as if some impending calamity awaited ; and should make the improvement of digestion the first business of life.

2 and 1. — Are like 3, only more so. Everything eaten gives pain, and life is but a burden.

To Cultivate. — Eat simple, plain, dry food, of which unbolted wheaten bread, and especially crackers made thereof, are best ; and but little at that, especially if the appetite is ravenous ; and masticate and salivate thoroughly ; eat in a cheerful, lively, pleasant spirit, talk-ing and laughing at meals ; consult appetite, or eat sparingly and leisurely that which relishes ; boiled wheat, or puddings made of wheaten flour, or grits, or oatmeal, or rye flour, eaten with cream and sugar, being the best staple article — say a teacupful of wheat or Gra-ham flour per day, thoroughly boiled ; should eat little after 5 P. M., and if hurried in business, before or after, but not during business hours, nor in a hurried, anxious state of mind, but as if determined to enjoy it ; above all, should cast off care, grief, business anxieties, troubles, and all painful remembrances and forebodings, and just lux-uriate in the passing moment.

Dyspepsia, now so alarmingly prevalent, is more a mental than corporeal disease — is consequent more on a worried, feverish, un happy state of mind, than stomachic disorder merely. It is usually brought on by eating very fast right after working very hard, and then working very hard right after eating too fast and too much, which allows so little energy to go to the stomach, that its contents ferment instead of being digested, which inflames the whole system, and causes morbid action in both the mental and physical functions. This inflam-mation creates a craving, hankering appetite, as well as a general irritable state of mind. But the more food is eaten the more it re-in-flames the stomach, and thereby re-increases these morbid hankerings ; while denying appetite diminishes this inflammation and consequent hungering and irritability. Sometimes eating gives temporary relief right before what has just been eaten ferments, but only re-increases the pain soon afterward. Starvation is the cure in all cases of a cra-ving appetite, but a poor appetite needs pampering, by providing any dainties that may relish. Or, perhaps the system is pining for want of some special aliment. If so, the appetite will hanker after it, and should be gratified, however seemingly unnatural, provided it be an alimentary article. (See Alimentiveness.) Above all, avoid alcohol and tobacco in all their forms, and also tea and coffee, using instead, a coffee made by browning wheat, rye, peas, corn, sweet potatoes, bread. etc., and prepared the same as Java.

Next, rub and pat, or lightly pound the stomach, liver, and bowels. While in college, a graduate came around advertising a specific panacea for dyspepsia, but requiring secrecy. It consisted simply in rubbing and kneading the abdomen. This supplies that mechanical action which restores them to functional action. Those manual exercises which call the abdominal muscles into special action, are preëminently useful, such as rowing, chopping wood, hoeing, and various gymnastic exercises.[1]

If the stomach is sore or painful, lay on at night a wet cloth, with a dry one over it, folded several thicknesses. If the bowels are torpid, induce an action of them at a given hour daily, and live much on boiled wheat, unbolted wheaten bread, and puddings, figs, and fruits, if the stomach will bear them. Observe all the health laws with scrupulous fidelity, relying more on nature, but little on medicines, and remit no efforts and spare no exertions to restore digestion ; for, till you do, you can only half think, study, remember, feel, transact business, or do or enjoy anything.

To RESTRAIN it, make less a god of the appetite, direct, or work up in other respects those energies now consumed by the stomach, and " be temperate in all things."

THE ABDOMINAL VISCERA complete the digestive functions. The stomach may solve its food, yet dormant liver, intestines, and mesentery glands fail to appropriate it. Or the latter may be good, and former poor.

7 and 6. — Are very fleshy, round-favored, and fat, and eliminate food material faster than it is consumed, besides sleeping well, and enjoying ease and comfort, and do only what must be done.

5 and 4. — Have a good, fair share of flesh and abdominal fullness, and appropriate about as much food as the system requires.

3. — Are rather slim, poor in flesh, and gaunt; may digest food well, but sluggish bowels and mesenteries fail to take up and empty into the circulation enough to fully sustain the life-functions, and have hence strong tendencies to constipation.

[1] See the author's new work on Physiology for the fullest exposition of this and all the other physical functions.

2. — Are very slim, poor, dormant, weak, and dyspeptic.

To RESTRAIN. — Breathe deeply, work hard, sleep little, **and eat** lightly.

THE MOTIVE OR MUSCULAR TEMPERAMENT.

Motion is a necessary and an integral part and parcel of life itself. What could man do, what be, without it? How walk, work, or move? How even breathe, digest, or circulate blood? — for what are these, indeed what all the physical functions, but action in its various phases?

And this action is effected by means of bones and muscles or fibres, the fleshy portions of the system. These bones constitute the founda- tion on which the muscular superstructure is built, are articulated at their ends by joints, and firmly bound together by ligaments which allow free motion. Toward the middle of these bones the muscles are firmly attached, so that when they contract they give motion to the end of the bone *opposite* the belly of the muscle. These muscles, of which there are some 527 in the human body, constitute the lean meat or red flesh of all animals, and are rendered red by the immense num- ber of minute blood-vessels which are ramified upon every fibre of every muscle, in order to resupply that vital power which is expended by its exercise. The contractile power of these muscles is truly aston- ishing, as is evident from the wonderful feats of strength and agility of which man is capable; and that, too, though these muscles act under a great mechanical disadvantage.

These bones and muscles collectively constitute the frame-work of the system — give it its build and form — are to the man what the timbers, ropes, and pulleys are to the ship, and constitute the Motive Temperament. Its predominance confers power of constitution, and strength of character and feeling.

7. — Are lean, spare; of good size and height, and athletic; have strongly marked features; a large, Roman nose; high and large cheek- bones; large and broad front-teeth; all the bones of the body project- ing; a deep, grum, bass voice; distinctly marked muscles and blood- vessels; large joints; hard flesh; great muscular power or physical strength; ease of action, and love of physical labor, of lifting, work- ing, etc.; dark, and often coarse, stiff, abundant, and perhaps bushy hair; if a man, a black and heavy beard; dark skin and eyes; a harsh, expressive visage; strong, but coarse and harsh feelings — the movements like those of the draught-horse, slow, but powerful and efficient; tough; thorough-going; forcible; strongly marked, if not

idiosyncratic; determined, and impressive both physically and mentally; and stamp their character on all they touch, of whom Alexander Campbell furnishes a good example. The motive, 7, mental, 6, and vital, 5, are capable of powerful and sustained mental effort, and great power in any department, especially that of mind as mind, of swaying a commanding influence o v e r mankind, taking the lead in a large business, etc. This temperament is always accompanied by prodigious coronal and perceptive regions, Firmness, and Combativeness, and large Destructiveness — its natural

MOTIVE, OR MUSCULAR TEMPERAMENT.

No. 131. — ALEXANDER CAMPBELL.

accompaniment — the very organs required to re-increase its force and efficiency, and indispensable to its exercise.

6. — Are like 7, except less in degree; are tough, hardy, and strong constitutioned; evince power, efficiency, and force in whatever is undertaken; use strong expressions; are stout, limber-jointed, and both need and can endure a world of action and fatigue; are like a fire made of anthracite coal, making a slow but powerful and continuous heat, and will make a decided mark in the business world, or in whatever other department these energies may be exercised. With the vital ·6 or 7, and the mental 3 or 4, are broad and prominent in form; large, tall, well proportioned, broad-shouldered, and muscular; usually coarse-featured, homely, stern, and awkward; enjoy hard work more than books or literary pursuits; have great power of feeling, and thus require much self-government; are endowed with good sense, but have a poor way of showing it; are strong-minded, but possess more talents than power to exhibit them; manifest talents more in managing machinery, creating resources, and directing large operations than in mind as such; improve with age, growing better and more intellectual

accomplish wonders; are hard to beat, indomitable, and usually useful citizens, but endowed with strong passions when once roused; and capable of being deeply depraved, especially if given to drink.

5. — Have a good share of the hearty, enduring, efficient, and potential; move right forward, with determination and vigor, irrespective of hindrances; bring a good deal to pass; and are like 6, only less so.

VITAL MOTIVE.

No. 132. — PHINEAS STEVENS.

4. — Are not deficient in motive power, yet more would be better; wrought up by special circumstances, can put forth unwonted strength; but it will be spasmodic, and liable to overstrain; can work hard, but are loth to; prefer the sedentary to the active, and business to labor; with the vital 6 or 7 are indolent physically, and do only what cannot be avoided, and need to cultivate muscular power.

3. — Dislike work; prefer sitting to moving, and riding to labor; may be quick and flashy, but are not powerful; lack strength and weight of character; *need* much more exercise than they love to take; and first of all should cultivate both muscular action and strength of character. With the vital 6, and mental 6 or 7, are rather small-boned, but plump, well formed, light complexioned, and often handsome; have usually auburn or flaxen hair; are most exquisitely organized, most pathetic and sympathetic, sentimental, exalted, and spiritual; have redoubled glow and fervor of feeling, derived from both the vital and mental, which they are hardly able to contain; easily receive and communicate impressions; are quite too much influenced by first impressions, and intuitive likes and dislikes; have hobbies; are most enthusiastic; throw a great amount of feeling into everything; use strong and hyperbolical expressions; are fond of company, if not forward in it; have a quick, clear, sharp, keen, active mind, and good business talents· a ready flow of ideas and a talent

for communicating them, either on paper or in social conversation; show taste, refinement, and delicacy in everything; have an under-current of pure, virtuous feeling, which will prevent the grosser manifestation of animal passion, and give the intellectual and moral the ascendency; sin only under some sudden and powerful excitement; are passionately fond of poetry, novels, tales, light and sentimental reading, belles-lettres, newspapers, etc., and inclined to attempt this kind of composition;

MOTIVE 3, MENTAL 7, VITAL 4.

No. 133. — FANNY FORESTER.

have a retentive memory, shrewdness, smartness, and enough of selfishness to take good care of self, yet not sufficient momentum or power to become great, but are rather effeminate. This temperament is found much oftener and more perfect in females than males, and is admirably illustrated by Fanny Forester. Children thus organized are precocious, and liable to die prematurely, and their physical culture would save to their parents and the world those brightest stars, which now generally set while rising, to shine no more on earth.

Mental 7, vital 5, and motive 3, may be smart, but cannot be great; may be brilliant, but are flashy, meteoric, vapid, too emotional, imaginative, and impulsive, and like a fire made of pine wood or shavings, intense, but momentary.

2 and 1. — Work, walk, move, and use muscles only when obliged to; pre-incline much more to the emotional and vapid than potential, and should cultivate the muscles assiduously.

MUSCULAR EXERCISE is indispensable to greatness and happiness. By a law of things, all parts must be exercised in about equal proportions. When the brain is worked more than the muscles, it becomes partially congested, loses its snap, leaves the mind dull, memory indefinite, and thought obtuse, which exercise remedies. None need ever think of becoming great intellectually, however splendid their heads or temperaments, without much vigorous exercise and real hard work, even. All eminent men have laid the foundations of their superiority by working hard during their minority, and continuing to exercise daily through life; while those students brought up without labor rarely take a high intellectual stand, except in parrot-like scholarship. They always lack vim and pith, and close, hard thought. And this deficiency grows on them. John Quincy Adams always rose before the sun to take his exercise, and as he became old took much of it in swimming, which he said gave the required exercise without heating his blood. Benton took a great amount of exercise. Jefferson always worked "like a Trojan." Webster would have his seasons of hunting, fishing, and rowing, besides taking a daily walk. Washington was a robust, hard-working farmer and soldier. *Physical* exercise is as indispensable to greatness as the intellectual organs themselves. And one principal reason why so many men, having all the phrenological indications of greatness, do not distinguish themselves, is a want of physical exercise.

To CULTIVATE. — Take all the muscular exercise you can well endure, but only gentle; make yourself comfortably tired every day; choose those kinds of exercise most agreeable, but practice *some* kind assiduously; dance more and sit less; if a child, should be allowed to run and play, to skate and slide down hill, romp and race, wrestle, practice gymnastics, climb and tear round all it likes, and furnished with playmates to encourage this out-of-door life. Fear neither exposure nor dirt, clothes or shoes, bad associates, or anything else which furnishes this great desideratum.

To RESTRAIN. — Use the muscles less and brain more.

7. — THE MENTAL TEMPERAMENT.

This embraces the brain and nerves, or that portion of the system called into exercise in the production of mind as such, or thought, feeling, sensation, memory, etc.

The brain consists at first of a mere ganglion of nervous matter, formed at the top of the spinal column. To this additions are made upward and forward, forming, successively, the brains of various animals, from that of the fish and toad, through that of the dog and monkey, up to the perfectly developed brain of the human adult. Let it be observed that the base of the brain, or the animal organs, which alone can be exercised by infants, are developed first, while Benevolence, Amativeness, Veneration, Constructiveness, and some others which cannot be exercised by them, are not developed till some years after birth.

The construction of the brain is most interesting. Its internal portion is fibrous, while its outer is soft and gelatinous. It is folded up into layers or furrows, called convolutions, which are expanded, by dropsy of the brain, into a nervous sheet or web. These convolutions allow a great amount of nervous matter to be packed up in a small compass, and their depth and size are proportionate to the amount of mind and talent. Thus in animals and idiots they are small and shallow; in men of ordinary talents much deeper; while the dissectors of the brains of Cuvier, Lord Byron, and other great men, remark with astonishment upon their size and depth.

Some writers say five times as much blood is sent to the brain in proportion to its volume as is sent to any other portion of the system, some say eight times, others fifteen, and one twenty; but all agree that it consumes many times more blood relatively, than any other part. The difference between them is doubtless owing to the difference in the talents of those experimented upon, intellectual subjects having the most. The distinctness and protrusion of the veins in the heads of great men, as also the immediate filling up of these veins when one laughs or becomes excited, have the same cause.

Through the medium of the spinal column, and by means of the nerves which go off from the spinal marrow through the joints of the back-bone, the brain holds intercourse with every part of the body, the nerves being ramified upon every portion of its surface, so that not even the point of a needle can penetrate any portion of it without lacerating them, and thus producing pain. This spinal marrow is composed of four principal columns, the two anterior ones exercising voluntary motion, the two posterior ones sensation. Let the nerves which go off from the two posterior columns be severed at their root, and the parts on which they are ramified will be destitute of sensation, not feeling anything, though able to move; but on severing the nerves which go off from the two anterior columns, though the patient will feel the prick of the needle, he will be unable to move the limb to which the nerve goes. Now, observe that these two *anterior* or motor columns are in direct connection with the *frontal* portion of the brain, in which the *intellectual* organs are located, so that each can communicate freely with the other, while the two *posterior* columns, or those of sensation, are in connection with the back part of the brain, in which the organs of the *feelings* are located. They are most abundant on the outer surface of the body, and accordingly the skin and adjacent flesh is the seat of much more intense pain from wounds than the internal portions.

7. — Have a small stature; light build; small bones and muscles; a slim, tall, spare, sprightly person; quickness of motion; great phys-

ical activity, too much for strength; sharp features and phrenolog ical organs; thin lips; small, pointed nose; and sharp teeth, liable to premature decay. [See Fanny Forester, cut 133.] Are characterized mentally by a predominance of mind over body, so that its states affect the body more than the body does the mind; are in the highest degree susceptible to the influence of stimuli, and of all exciting causes; are refined and delicate in feeling and expression, and easily disgusted with anything coarse, vulgar, or out of taste; enjoy and suffer in the highest degree; are subject to extremes of feeling; have the disgusts, sympathies, and prepossessions easily excited; experience a vividness and intensity of emotion, and a clearness, pointedness, and rapidity of thought, perception, and conception, and a love of mental exercise imparted by no other temperament; have a deep flow of pure and virtuous feeling, which will effectually resist vicious inclinations; intense desires, and put forth correspondingly vigorous efforts to gratify them; are eager in pursuits, and feel that their ends are of the utmost importance, and must be answered *now;* are thus liable to overdo, and prematurely exhaust the physical powers, which are poor at best; are very fond of reading and study, of thinking and reasoning, of books and literary pursuits, of conversation, and all kinds of information, and apt to lie awake at night, thinking, or feeling, or reading; incline to some profession, or light mental occupation, such as a clerk, merchant, teacher, or if a mechanic, should be a goldsmith, or architect, or something requiring light action, but not hard lifting, or more head work than hand work; should avoid close application; take much pleasurable recreation and exercise; avoid all kinds of stimulants, wines, tobacco, tea and coffee included; endeavor to enjoy existence; and avoid being worried.

6. — Are like 7 in character, only less in degree; more given to intellectual and moral than animal pleasures, and action than rest; cannot endure slow or stupid employees; with the motive 6, are of good size; rather tall, slim, lean, and raw-boned, if not homely and awkward; have prominent bones and features, particularly front teeth and nose; a firm and distinct muscle; a tough, wiry, excellent physical organization; a firm, straightforward, rapid, energetic walk; great ease and efficiency of action, with little fatigue; a keen, penetrating eye; large joints, hands, feet, etc.; a long face and head, and a high head and forehead; a brain developed more from the nose over to the occiput than around the ears; large intellectual and moral

organs; strong desires, and great power of will and energy of character; vigorous passions; a natural love of hard work, and capacity
for carrying forward and managing great undertakings; that thorough-going spirit which takes right hold of great projects with both
hands, and drives into and through thick and thin, in spite of obstacles
and opposition, however great, and thus accomplishes wonders; superior business talents; unusual strength and vigor of intellect; strong
common sense; good general judgment; with a large intellectual
lobe, and a cool, clear, long, calculating head; a reflective, planning,
discriminating cast of mind, and talents more solid than brilliant;
are more fond of the natural sciences than literature; of philosophy
than history; of the deep, solid branches than belles-lettres; of a
professional and mental than laborious vocation; of mental than
bodily action; and the moral than sensual.

5. — Have good, fair muscles; are quite prominent-featured, easy
of motion, enduring, tough, hardy, clear-headed, and fond of intellectual pursuits; have good ideas, and excellent native sense and
judgment; talk, speak, and write to the purpose, if at all; love action and exercise, and walk and work easily; are efficient, and capable of doing up a good life labor, but not a genius. With the vital
6, are sprightly, lively, vivacious, and happy; and with the motive 3,
are not adapted to a life of labor, but should choose some office business, yet exercise a great deal — no matter how much.

4. — Have fair mental action, if circumstances fully call it forth;
if not, are commonplace; must depend for talents more on culture
and plodding studiousness than natural genius; with culture, can do
well, without it little; with the motive and vital 6 or 7, are far better adapted to farming or manual pursuits than literary, and should
cultivate intellect and memory.

3. — Have little love of literary pursuits; are rather dull, and fall
asleep over books and sermons; and cannot marshal ideas for speaking or writing.

2. — Are exceedingly dull of comprehension; slow of perception;
poor in judgment and memory; hate books; must be told what and
how to do; and should seek the direction of superior minds.

1. — Are almost senseless and idiotic.

A WELL-BALANCED TEMPERAMENT

Is by far the best, that most favorable to true greatness and general

A WELL-BALANCED TEMPERAMENT.

No. 134. — WASHINGTON.

genius, to strength of character, along with perfection, and to harmony and consistency throughout, is one in which each is strongly marked, and all about equally developed.

Excessive motive with deficient mental gives power with sluggishness, so that the talents lie dormant. Excessive vital gives physical power and enjoyment, but too little of the mental and moral, along with coarseness and animality. Excessive mental confers too much mind for body, too much sentimentalism and exquisiteness, along with greenhouse precocity. Whereas their equal balance gives an abundant supply of vital energy, physical stamina, and mental power and susceptibility. They may be compared to the several parts of a steamboat and its appurtenances. The vital is the steam-power; the motive, the hulk or frame-work; the mental, the freight and passengers. The vital predominant generates more animal energy than can well be worked off, which causes restlessness, excessive passion, and a pressure which endangers outbursts and overt actions; predominant motive gives too much frame or hulk; moves slowly, and with weak mental, is too light freighted to secure the great ends of life; predominant mental overloads, and endangers sinking; but all equally balanced and powerful, carry great loads rapidly and well, and accomplish wonders. Such persons unite cool judgment with intense and well-governed feelings; great force of character and intellect with perfect consistency; scholarship with sound common sense; far-seeing sagacity with brilliancy; and have the highest order of both physiology and mentality. Such a temperament had he immortal Washington, and his character corresponded.

Most diseases, too, are consequent on this predominance or deficiency of one or another of these temperaments, and when either fail, all fail. Hence the infinite importance of cultivating those that are weak. A well-balanced phrenology is equally important, and its absence unfavorable.

7 or 6. — Are uniform, consistent, harmonious in character, even-tempered, popular, and generally liked; not remarkable for any specialties of talents or character, nor for any deficiencies, and "maintain the even tenor of their way" among men.

5 or 4. — Are in the main consistent, and in harmony with them-selves, but more or less affected by circumstances; show general uniformity of life and doctrine, but different circumstances change their characters.

3. — Have uneven heads and characters; are singular in expression, looks, and doctrine, and variable in conduct; often inconsistent, and with excitability 6 or 7, the creatures of circumstances; take one-sided views of things; are poor counselors; need and should take advice; are easily warped in judgment; propound strange ideas, and run after novelties; and need to cultivate unity and homoge-neousness of opinion and conduct.

2. — Are like 3, only more so; are nondescripts; idiosyncratic in everything; just like themselves, but unlike anybody else; and nei-ther like, nor are liked by, others.

To CULTIVATE. — Exercise the weaker and restrain the stronger faculties and temperaments according to directions in this work.

HOMOGENEOUSNESS, OR ONENESS OF STRUCTURE.

Every part of everything bears an exact correspondence to every other part of it. Thus, tall-bodied trees have long branches and leaves; short-bodied trees, short branches and roots; and creeping vines, as the grape, honeysuckle, etc., long, slim roots that run under ground as extensively as their tops do above. The Rhode Island Greening, a large, well-proportioned apple, grows on a tree large in trunk, limb, leaf, and root, and symmetrical, while the gillefleur is conical, and its tree long-limbed, and runs up high to a peak at the top, while flat and broad-topped trees bear wide, flat, sunken-eyed apples. Very thrifty growing trees, as the Baldwin, Fall Pippin, Bartlett, Black Tartarian, etc., generally bear large fruit; while small fruit, as the Seckel pear, Lady Apple, Belle de Choisy cherry, etc., grow slowly and have many small twigs and branches. Trees that bear red fruit, as the Baldwin, etc., have red inner bark; while yellow and green-colored fruits grow on trees the inner rind of whose limbs is yellow or green. Peach-trees that bear early peaches have deeply-notched leaves, and the converse of late ones; so that, by these and other physiognomical signs, experienced nurserymen can tell what a given tree bears at first sight.

Correspondingly, long-handed persons have long fingers, toes, arms, legs, bodies, heads, and phrenological organs; while short and broad-shouldered persons are short and broad-handed, fingered, faced, nosed, and limbed, and wide and low bodied. When the bones on the hand are prominent, all the bones, nose included, are equally so, and thus of all other characteristics of the hand, and every other portion of all bodies. Hence, every hand proclaims the general character of its owner, because if it is large or small, hard or soft, strong or weak, firm or flabby, coarse-grained or fine-textured, even or prominent, rough or smooth, small-boned or large-boned, or whatever else, the whole body is built upon the same principle, with which the brain and mentality also correspond. Hence, also, small-nosed persons have little soul, and large-nosed a great deal of character of some kind.

Bonaparte chose large-nosed men for his generals, and the opinion prevails that large noses indicate long heads and strong minds, not because great noses cause great minds, but because the motive or powerful temperament causes both. Flat noses indicate flatness of mind and character, by indicating a poor, low, organic structure. Broad noses indicate large passage-ways to the lungs, and therefore large lungs and vital organs, and this, great strength of constitution, and hearty animal passions, along with selfishness; for broad noses, broad shoulders, broad heads, and large animal organs go together. But when the nose is narrow at the base, the nostrils are small, because the lungs are small, and need but small avenues for air, which indicates a predisposition to consumptive complaints, along with an active brain and nervous system, and a passionate fondness for literary pursuits. Sharp noses indicate a quick, clear, penetrating, searching, knowing, sagacious mind, and also a scold; indicate warmth of love, hate, generosity, moral sentiment — indeed, positiveness in everything, while blunt noses indicate and accompany obtuse intellects and perceptions, sluggish feelings, and a soulless character. The Roman nose indicates a martial spirit, love of debate, resistance, and strong passions, while straight, finely-formed Grecian noses indicate harmonious characters. Seek their acquaintance. We have chosen our illustrations from the nose, because it is easily seen and

described, and renders observations on the character easy and correct. But the *principle* here exemplified applies to all the other
organs and portions of the face and body.

And the general forms of the head correspond with those of the
body and nose. Where the nose is sharp, all the bones and phrenological organs, and of course mental characteristics, are equally
sharp — the whole person being built on the sharp principle, and
thus of breadth, prominence, length, etc.

Tall persons have high heads, and are aspiring, aim high, and seek
conspicuosity, while short ones have flat heads, and seek the lower
forms of worldly pleasures. Tall persons are rarely mean, though
often grasping; but very penurious persons are often broad-built.
Small persons generally have exquisite mentalities, yet less power —
the more precious the article the smaller the package in which it is
done up — while great men are rarely dwarfs, though great size often
coexists with sluggishness. To particularize — there are four leading
forms which indicate generic characteristics, all existing in every one,
yet in different DEGREES. They are —

1. — PROMINENCE INDICATES POWER.

" A lean horse for a long pull " is an observation as true as trite.
This corresponds with the motive temperament, which it indicates.

2. — BREADTH AS INDICATING ANIMALITY.

Spherical forms are naturally self-protecting. Roundness protects
its possessor. So all round-built animals, as Indian pony, bull-dog,
elephant, etc., are strong-constitutioned, tough, enduring, and very
hardy, but less active and sprightly in body and mind. And this
applies equally to human beings. Broad-built persons may be industrious, plodding, good-feeling, and the like, but love their ease, are
not brilliant, and take good care of self. Yet they wear like iron,
and unless health has been abused, can live to a great age. This
form corresponds with the vital temperament.

3. — ACTIVITY INDICATED BY LENGTH.

In and by the nature of things length of form facilitates ease of
action. Thus, deer, gazelle, greyhound, giraffe, tiger, weasel, ermine,
eel, and all long and slim animals, are quick-motioned, lively,
sprightly, nimble, and agile. The same principle applies equally to
persons. Thus, those very long-favored, or in whom this form is

7. —· Are as quick as a flash to perceive and do; agile; light-motioned, limber-jointed; nimble; always in motion; restless as the wind; talk too rapidly to be emphatic; have no lazy bones in their bodies; are always moving head, hands, feet, something; are natural scholars; quick to learn and understand; remarkably smart and knowing; loving action for its own sake; wide awake; eager, uncommonly quick to think and feel; sprightly in conversation; versatile in talent; flexible; suggestive; abounding in idea; apt at most things; predisposed to consumption, because action exceeds strength; early ripe; brilliant; liable to premature exhaustion and disease, because the mentality predominates over the vitality, of which the late Captain Knight, who had a world-wide reputation for activity. enterprise, daring, impetuosity, promptness, judgment, earnestness, executiveness, affability, and sprightliness, furnishes a good example.

LONG, SHARP, AND ACTIVE.

No. 135. — CAPTAIN E. KNIGHT.

6. — Are active, restless, brisk, stirring, lively, anything but lazy, with a good organism; are quick-spoken; clear-headed; understand matters and things at the first glance; see right into and through business, and all they touch, readily; are real workers with head or hands, but prefer head-work; positive; the one thing or the other; and are strongly pre-inclined to the intellectual and moral. Their characters, unless perverted, like their persons, ascend instead of descending; and they are better adapted to law, merchandise, banking, or business than to farming, or heavy mechanical work. Yet, if mechanics, should choose those kinds requiring more sprightliness than strength, and mind than muscle.

5 or 4. — Have a fair, but only fair, share of natural activity and sprightliness; do what they well can, and with tolerable ease, but do not love action for its own sake.

3. — Are rather inactive; do only what they must, and that grudg-

ingly; love to be waited on, but not to wait; and get along with the fewest steps possible; seek a sedentary life, and are as loth to exercise mind as body.

2 and 1. — Are downright slothful, lazy, and good for nothing to themselves or others.

To CULTIVATE. — Keep doing, doing, doing all the time, and in as lively and sprightly a manner as possible; and live more on foot than seated.

To RESTRAIN. — Sit down and rest when tired, and let the world jog on while you enjoy it. Do only half you think you must, and be content to let the rest go undone. Try to be lazy. Work as few hours as possible, and get along with the least outlay of strength possible. Do sit down, and enjoy what you have already got, instead of trying to get so much more. Live on your laurels. Don't tear and fret if all is not exactly to your liking, but cultivate contentment.

4. — EXCITABILITY INDICATED BY SHARPNESS.

All sharp things are, in and by the very nature of their form, penetrating, of which the needle furnishes an example. And this law applies equally to human beings. From time immemorial a sharp nose has been considered indicative of a scolding disposition; yet it is equally so of intensity in the other feelings, as well as temper.

7. —Are extremely susceptible to impressions of all kinds; intensely excited by trifles; apt to magnify good, bad, everything, far beyond the reality; a creature of impulse and mere feeling; subject to extreme ups and downs of emotion; one hour in the garret, the next in the cellar; extremely liable to neuralgia and nervous affections; with quality and activity 6 or 7, have ardent desires; intense feelings; keen susceptibilities; enjoy and suffer in the extreme; are wholesouled; sensitive; positive in likes and dislikes; cordial; enthusiastic; impulsive; have hobbies; abound in good-feeling, yet are quick-tempered; excitable; liable to extremes; have a great deal of SOUL or passion, and warmth of feeling; are BRILLIANT writers or speakers, but too refined and sensitive for the mass of mankind; gleam in the career of genius, but burn out the vital powers on the altar of nervous excitability, and like Pollok, Henry Kirke White, Macdonald Clarke, and Leggett, fall victims to premature death, and should keep clear from all false excitements and stimulants, mental and physical, such as tea, coffee, tobacco, drugs, and alcoholic drinks, and cool off and keep cool.

6. — Are like 7, only less so; warm-hearted, impetuous, impulsive, full of soul, and too susceptible to external influences; swayed too much by feeling; and need much self-government and coolness.

5. — Are sufficiently sensitive and susceptible to exciting causes, yet not passional, nor impulsive; and easily roused, yet not easily carried away by excitements. With activity 6 or 7, are very quick, but perfectly cool; decide and act instantly, yet knowingly; do nothing without thinking, but think and do instantaneously; are never flustered, but combine great rapidity with perfect self-possession.

4. — Are like the placid lake — no waves, no noise, and evince the same quiet spirit under all circumstances.

3. — Are rather phlegmatic; slow to perceive and feel; rather cold and passionless; rarely ever elated or depressed; neither love nor hate, enjoy nor suffer, much; are enthusiastic in nothing, and throw little life or soul into expressions or actions.

2. — Are torpid, soulless, listless, spiritless, half asleep about everything, and monotonous and mechanical in everything.

1. — Are really stupid, and about as dead and hard as sole-leather — having the texture of humanity, but lacking its life and glow, and enjoy and suffer very little.

To CULTIVATE. — Yield yourself up to the effects or influences of persons and things operating on you; seek amusements and excitements; and try to feel more than comes natural to you.

To RESTRAIN. — First, fulfill all the health conditions, so as thereby to allay all false excitement, and secure a quiet state of the body. Eat freely of lettuce, but avoid spices and condiments. Air, exercise, water, and sleep, and avoiding stimulants, constitute your great physical opiates. Second, avoid all unpleasant mental excitements, and by mere force of will cultivate a calm, quiet, luxurious, to-day-enjoying frame of mind. If in trouble, banish it, and make yourself as happy as possible. Take lessons of Quakers.

These primary forms and characteristics usually combine in different degrees, producing, of course, corresponding differences in the talents and characteristics. Thus, eloquence accompanies breadth combined with sharpness. They create that gushing sympathy, that spontaneous overflowing of soul, that high-wrought, impassioned ecstasy and intensity of emotion in which true eloquence consists, and transmit it less by words than look, gesture, and those touching, melting, soul-stirring, thrilling intonations which storm the citadel of the soul. Hence it can never be written, but must be seen, heard, and felt. This sharpness and breadth produce it first by giving great lungs to exhilarate the speaker, and send the blood frothing and foaming to the brain, and secondly, by conferring the utmost excitability and intensity of emotion; and it is in this exhilaration that real eloquence mainly consists. This sharp and broad form predominates in Bascom, whom Clay pronounced the greatest natural orator he ever heard; in Chapin and Beecher, to-day confessedly our finest speakers in the pulpit or the rostrum; in Everett; in " the old man eloquent," indeed, both the Adamses; in Dr. Bethune and a host of others. Still, in Patrick Henry, Pitt, and John B. Gough, each unequaled in his day and sphere, the sharp combines with the long. This gives activity united with excitability. Yet this form gives also the poetic more than the oratorical — gives the impassioned, which is the soul of both.

Authorship, again, is usually accompanied by the long, prominent, and sharp. Reference is not now had to flippant scribblers of exciting newspaper squibs, or even of dashing editorials, or highfalutin productions, nor to mere compilers, but to the authors of deep, sound, original, philosophical, clear-headed, labored productions. It predominates in Revs. Jonathan Edwards, Wilbur Fiske, N. Taylor, Dr. E. A. Parke, Leonard Bacon, Albert Barnes, Oberlin, Pres. Day, Drs. Parish and Rush, Pres. Hitchcock, Hugh L. White, Dr. Caldwell, Elias Hicks, Franklin, Alexander Hamilton, Chief-Justice Marshall, Calhoun, John Q. Adams, Percival, Noah Webster, George Combe, Lucretia Mott, Catherine Waterman, Mrs. Sigourney, and nearly every distinguished author and scholar.

THE POETIC, OR LONG AND SHARP FORM.

Poetry inheres in various forms. Some distinguished poets are broad and sharp, others long and sharp, but all sharp. Those who evolve the highest, finest, and most fervid style and cast of sentiment, have more of the long, with less of the prominent, yet with the long predominating over the sharp, and are often quite tall. Wm. C. Bryant furnishes an excellent illustration of this shape, as his character does of its accompanying mentality. Those who poetize the passions are, like orators, broad and sharp, of whom Byron furnishes an example in poetry and configuration. The best combination of forms for writers and scholars is the sharp predominant, long next, prominent next, and all conspicuous. The best form for contractors, build-

4

THE MENTAL-MOTIVE TEMPERAMENT.

No. 136. — WILLIAM CULLEN BRYANT.

ers, managers of men and large mechanical operations, is the broad and prominent combined. (Cut 132.) But they should not be slim. A farmer may have any form but a spindling one, yet a horticulturist or nurseryman may be slim.

RESEMBLANCE BETWEEN MEN AND ANIMALS.

That certain men "look like" one or another species of animals is an ancient observation. And when in looks, also in character. That is, some have both the lion, or bull-dog, or eagle, or squirrel expression of face, and likewise traits of character. Thus, Daniel Webster was called the "Lion of the North," from his general resemblance in form, heavy shoulders, hair, and general expression to that king of beasts; and a lion he indeed was, in his sluggishness when at his ease, but power when roused; in his magnanimity to opponents, and the power of his passions.

He had a distinguished contemporary, whose color, expression of countenance, manners, everything, resembled those of the fox, and was he not foxy in character as well as looks? And did he not introduce into the political machinery of our country that wire-working, double-game policy and chicanery, which has done more to corrupt our ever-glorious institutions than everything else combined, even endangering their very existence? Freemen, vote only for open-handed, honest men; never for tricksters.

No. 137. — DANIEL WEBSTER. — THE LION FACE.

Those who resemble the bull-dog are broad-built, round favored square-faced, round-headed, having a forehead square, and perhaps prominent, but low; mouth rendered square by the projection of the eye or canine teeth, and smallness of those in front; corners of the mouth drawn down; and voice deep, guttural, growling, and snarling. Such, if fed, will bark and bite *for* you, but, if provoked, will lay right hold of you, and hold on till you or they perish in the desperate struggle. And when this form is found on female shoulders, " the Lord deliver you."

Tristam Burges, called in Congress the "Bald Eagle," from his having the aquiline or eagle-bill nose, a projection in the upper lip, falling into an indentation in the lower, his eagle-shaped eyes and eye-brows, as seen in the accompanying engraving, was eagle-like in character, and the most sarcastic, tearing, and soaring man of his day, John Randolph excepted. And whoever has a long, hooked, hawk-bill, or Roman nose, wide mouth, spare form, prominence at the lower and middle part of the forehead, is very fierce when assailed, high-tempered, vindictive, efficient, and aspiring, and will

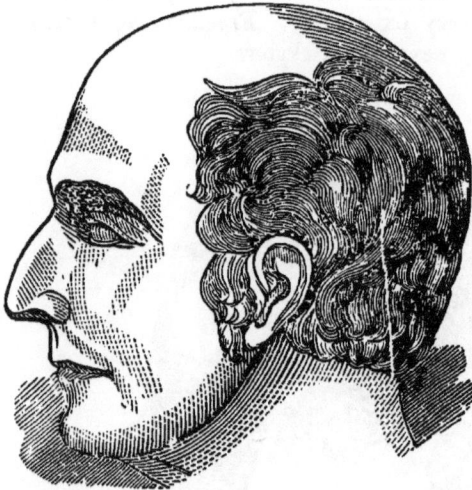

No. 138. — Tristam Burges. — The Eagle.

fly higher and farther than others.

Tigers are always spare, muscular, long, full over the eyes, large-mouthed, and have eyes slanting downward from their outer to inner angles; and human beings thus physiognomically characterized, are fierce, domineering, revengeful, most enterprising, not over humane, a terror to enemies, and conspicuous somewhere.

Swine — fat, loggy, lazy, good-dispositioned, flat and hollow-nosed — have their cousins in large-abdomened, pug-nosed, double-chinned, talkative, story-enjoying, beer-loving, good-feeling and feeding, yes-yes humans, who love some easy business, but hate HARD work.

Horses, oxen, sheep, owls, doves, snakes, and even frogs, etc., also have their men and women cousins, with their accompanying characters.

These resemblances are more easily seen than described; but the voice, forms of mouth, nose, and chin are the best bases for observation.

BEAUTIFUL, HOMELY, AND OTHER FORMS.

In accordance with this general law, that shape is as character, well-proportioned persons have harmony of features and well-balanced minds; whereas those, some of whose features stand right out, and others fall far in, have uneven, ill-balanced characters, so that homely, disjointed exteriors indicate corresponding interiors, while evenly-balanced and exquisitely formed men and women have well-balanced and susceptible mentalities. Hence, woman, more beautiful than man, has finer feelings and greater perfection of character, yet is less powerful — and the more beautifully formed the more exquisite and perfect the mentality. Nature never deceives — never clothes that in a beautiful, attractive exterior which is intrinsically bad or repellant. True, the handsomest women sometimes make the greatest scolds, just as the sweetest things when soured become correspondingly sour. The finest things, when perverted, become the worst. Those naturally beautiful and exquisitely organized, when perverted become proportionally bad, and those naturally ugly-formed are naturally bad-dispositioned.

Yet homely persons are often excellent tempered, benevolent, talented, etc., because they have a few POWERFUL traits, and also features — the very thing we are describing — that is, they have EXTREMES alike of face and character. Thus it is that every diversity of character has its correspondence in both the physiognomical form and organic texture.

INTONATIONS AS EXPRESSING CHARACTER.

Whatever makes a noise, from the deafening roar of sea, cataract, and whirlwind's mighty crash, through all forms of animal life, to the sweet and gentle voice of woman, makes a sound which agrees perfectly with the maker's character. Thus the terrific roar of the lion, and the soft cooing of the dove, correspond exactly with their respective dispositions; while the rough and powerful bellow of the bull, the fierce yell of the tiger, the coarse, guttural moan of the hyena, the swinish grunt, the sweet warblings of birds, in contrast with the raven's croak and owl's hoot, all correspond perfectly with their respective characteristics. And this law holds equally true of man. Hence human intonations are as superior to brute as human character exceeds animal. Accordingly, the peculiarities of all human beings are expressed in their voices and mode of speaking. Coarse-grained and powerful animal organizations have a coarse, harsh, and grating voice, while in exact proportion as persons become refined and elevated mentally, will their tones of voice become correspondingly refined and perfected. We little realize how much character we infer from this source. Thus, some female friends are visiting me transiently. A male friend, staying with me, enters the room, is seen by my female company, and his walk, dress, manners, etc., closely scrutinized, yet he says nothing, and retires, leaving a comparatively indistinct impression as to his character upon my female visitors, whereas, if he simply said yes or no, the mere SOUND of his voice communicates to their minds much of his character, and serves to fix distinctly upon their minds clear and correct general ideas of his mentality.

The barbarous races use the guttural sounds more than the civilized. Thus Indians talk more down the throat than white men, and thus of all, whether lower or higher in the human scale. Those whose voices are clear and distinct have clear minds, while those who only half form their words, or are heard indistinctly, say by deaf persons, are mentally obtuse. Those who have sharp, shrill intonations have correspondingly intense feelings and equal sharpness both of anger and kindness, as is exemplified by every scold in the world; whereas those with smooth or sweet voices have corresponding evenness and goodness of character. Yet, contradictory as it may seem, these same persons not unfrequently combine both sharpness and softness of voice, and such always combine them in character. There are also the intellectual, the moral, the animal, the selfish, the benignant, the mirthful, the devout, the loving, and many other intonations, each accompanying corresponding peculiarities of characters. In short, every individual is compelled, by every word uttered, to manifest something of the true character — a sign of character as diversified as correct.

COLOR AND TEXTURE OF HAIR, SKIN, BEARD, ETC.

Everything in nature is colored, inside and out; and the color always corresponds with the character. Nature paints her coarse productions in coarse drab, but adorns all her finer, more exquisite productions with her most beautiful colors. Thus, highly-colored fruits are always highly-flavored, and birds of the highest quality are arrayed in the most gorgeous tints and hues.

So, also, particular colors signify particular qualities. Thus, throughout all nature *black* signifies power, or a great amount of its characteristics; *red*, the ardent, loving, intense, concentrated, positive; *green*, immaturity; *yellow*, ripeness, richness, etc. Hence all black animals are powerful, of which the bear, Morgan horse, black snake, etc., furnish examples. So black fruits, as blackberry, black raspberry, whortleberry, black tartarian cherry, etc., are highly flavored and full of rich juices. So, also, the dark races, as Indian and African, are strong, muscular, and very tough. All red fruits are acid, as the strawberry; but the darker they are the sweeter, as the baldwin, gillifleur, etc.; while striped apples blend the sweet with the sour. Whatever is growing, or still immature, is green; but all grasses, grains, fruits, etc., pass, while ripening, from the green to the yellow, and sometimes through the red. Fruits red and yellow are always delicious. Other primary colors signify other characteristics.

Now, since coarseness and fineness of texture indicate coarse and fine-grained feelings and characters, and since black signifies power, and red ardor, therefore coarse black hair and skin signify great power of character of some kind, along with considerable tendency to the sensual; yet fine black hair and skin indicate strength of character, along with purity and goodness. Dark-skinned nations are always behind the light-skinned in all the improvements of the age, as well as in the higher and finer manifestations of humanity. So, too, dark-haired persons, like Webster, sometimes called "Black Dan," possess great power of intellect and propensity, yet lack the finer and more delicate shadings of sensibility and purity. Coarse black hair and skin, and coarse red hair and whiskers, indicate powerful animal passions, together with corresponding strength of character; while fine, light, and auburn hair indicate quick susceptibilities together with refinement and good taste. Fine dark or brown hair indicates the combination of exquisite susceptibilities with great strength of character, while auburn hair and a florid countenance, indicate the highest order of sentiment and intensity of feeling, along with corresponding purity of character, combined with the highest capacities for enjoyment and suffering. And the intermediate colors and textures indicate intermediate mentalities. Curly hair and beard indicate a crisp, excitable, and variable disposition, with much diversity of character — now blowing hot, now cold — along with intense love and hate, gushing, glowing emotions, brilliancy, and variety of talent. So look out for ringlets; they betoken April weather. Treat them gently, lovingly, and you will have the brightest, clearest sunshine, and the sweetest, balmiest reezes; but ruffle them, and you raise a storm, a very hurricane, changeable, now so very hot, now so cold. Better not ruffle them. And this is doubly true of auburn curls; though auburn ringlets

need but a little right, kind, fond treatment to render them all as **fair**
and delightful as the brightest spring morning.

Straight, even, smooth, and glossy hair indicates strength, harmony,
and evenness of character, and hearty, whole-souled affections, as
well as a clear head and superior talents ; while stiff, straight, black
hair and beard indicate a coarse, strong, rigid, straightforward character. Abundance of hair and beard signifies virility and a great
amount of character.

Coarse-haired persons should never turn dentists or clerks, but
seek some out-door employment, and would be better contented with
rough, hard work than a light or sedentary occupation, although
mental and sprightly occupations would serve to refine and improve
them ; while dark and fine-haired persons may choose purely intellectual occupations, and become lecturers or writers with fair prospects
of success. Red-haired persons should seek out-door employment,
for they require a great amount of air and exercise ; while those who
have light, fine hair should choose occupations involving taste and
mental acumen, yet take bodily exercise enough to tone up and invig-
orate their system.

Generally, when either skin, hair, or features are fine or coarse, the
others are equally so. Yet some inherit fineness from one parent,
and coarseness from the other, while the color of the eye generally
corresponds with that of the skin, and expresses character. Ligh
eyes indicate warmth of feeling, and dark eyes power.

The mere expression of the eye conveys precise ideas of the existing and predominant states of the mentality and physiology. As
long as the constitution remains unimpaired, the eye is clear and
bright, but becomes languid and soulless in proportion as the brain
has been enfeebled. Wild, erratic persons have a half crazed expression of eye, while calmness, benignancy, intelligence, purity, sweetness, love, sensuality, anger, and all the other mental affections,
express themselves quite as distinctly by the eye as voice, or any
other mode.

REDNESS AND PALENESS OF FACE.

Thus far our remarks have appertained to the constant colors of
the face, yet those colors are often diversified or changed for the
time being.

Thus, at one time the whole countenance will be pale, at another
very red ; each of which indicates the existing states of body and

mind. Or thus : when the system is in a perfectly healthy state, the whole face will be suffused with the glow of health and beauty, and have a red, but never an inflamed aspect ; but any permanent injury of health, which prostrates the bodily energies, will change this florid complexion into dullness of countenance, indicating that but little blood comes to the surface or flows to the head, and a corresponding stagnation of the physical and mental powers. Yet, after a time, this dullness frequently gives way to a fiery redness ; not the floridness of health, but the redness of inflammation or false excitement; which indicates a corresponding depreciation of the mental faculties. Dark or livid red faces, so far from signifying the most health, frequently betoken the most disease, and correspondingly more animal and sensual characters ; because physiological inflammation irritates the propensities more, relatively, than the moral and intellectual faculties, though it increases the latter also. When the moral and intellectual faculties greatly predominate over the animal, redness may not cause coarse animality, because, while it heightens the animal nature, it also increases the intellectual and moral, which, being the larger, hold them in check ; but when the animal about equals or exceeds the moral and intellectual, this inflammation evinces a greater increase of animality than intellectuality and morality. Gross sensualists and depraved sinners generally have a fiery red countenance. Stand aloof from them, for their passions are all on fire, ready to ignite and explode on provocations so slight that a healthy physiology would scarcely notice them. This point can hardly be made fully intelligible ; but let readers note the difference between a healthy floridness of face and the fiery redness of drunkards, debauchees, etc. Nor does an inflamed physiology increase the animal nature, merely ; it also gives it a far more depraved and sensual cast, thereby doubly increasing the depraved tendencies.

PHRENOLOGICAL SIGNS OF CHARACTER AND TALENTS.

All truth bears upon its front unmistakable evidence of its divine origin, in its philosophical beauty, fitness, and consistency ; whereas, all untruth is grossly and palpably deformed. Any truth, also, harmonizes with all other truth, and conflicts with all error, so that, to ascertain what is true, and detect what is false, is perfectly easy. Apply this test, intellectual reader, to one after another of the doctrines taught by Phrenology.

The brain is both the organ of the mind, the dome of thought, the palace of the soul, and equally the organ of the *body*, over which it exerts an all-potent influence for good or ill, to weaken or stimulate, to kill or make alive. In short, the brain is the organ of the body in general, and of each of its organs in particular. As the stomach has its cerebral organ in Alimentiveness, and the muscular system its in Muscularity, so undoubtedly the lungs, liver, pancreas, bowels, etc., have each its yet undiscovered cerebral organ located in the under side of the brain. It sends forth those nervous energies which keep muscles, liver, bowels, and all the other bodily organs in a high or low state of action; and, more than all other causes, invites or repels disease, prolongs or shortens life, and treats the body as its galley-slave. Hence, healthy cerebral action is indispensable to bodily health, while a longing, pining, dissatisfied, fretful, or troubled state of mind is most destructive of health, and productive of disease. So is violence in any and all the passions. Indeed, the state of the mind is mainly controlled by that of the health. Even dyspepsia is more a mental than physical condition, and to be cured first and mainly by banishing that agitated, flashy, eager, craving state of mind, and securing instead a calm, quiet, let-the-world-slide state; nor will any physical appliances avail much without this mental restorative. Hence, too, we walk or work so much more easily and efficiently when we take an *interest* in what we do. Therefore, those who would be happy or talented must first and mainly keep their BRAIN vigorous and healthy.

The brain is subdivided into two hemispheres, the right and left, by the falciform process of the dura mater — a membrane which dips down one to two inches into the brain, and runs from the root of the nose over to the nape of the neck. This arrangement renders all the phrenological organs DOUBLE. Thus, as there are two eyes, ears, etc., in order that when one is diseased, the other can carry forward the function, so there are two lobes to each phrenological organ, one on each side.

The brain is divided thus. The feelings occupy that portion commonly covered by hair, while the forehead is occupied by the intellectual organs. These greater divisions are subdivided into the animal brain, located between and around the ears; the aspiring faculties, which occupy the crown of the head; the moral and religious sentiments, which occupy its top; the physico-perceptives, located over the eyes; and the reflectives, in the upper portion of the fore head. The predominance of each of these respective groups produces both particular shapes of head, and corresponding traits of character. Thus, a head projecting far back behind the ears, and hanging over and downward in the occipital region, indicates very strong domestic ties and social affections, a love of home, its relations and endearments, and a corresponding capacity of being happy in the family, and making family happy. The social affections are located in the *back* part of the head; and, accordingly, woman being more loving than man, when not under the influence of the other faculties, usually inclines her head backward; and when she kisses children, and those she loves, always turns the head directly backward, and rolls it from side to side, on the back of the neck. Wide and round heads, on the contrary, indicate strong animal and selfish propensities, while thin, narrow heads indicate a corresponding want of selfishness and animality. A head projecting far up at the crown indicates an aspiring, self-elevating disposition, pride of character, and a desire to be and to do something great; while a flattened crown indicates a want of ambition, energy, and aspiration. A head high, long, and wide upon the top, but narrow between the ears, indicates Causality, moral virtue, much practical goodness, and a corresponding elevation of character; while a low and narrow top-head indicates a corresponding deficiency of these humane and religious susceptibilities. A head wide at the upper part of the temples indicates a corresponding desire for personal perfection, together with a love of the beautiful and refined, while narrowness in this region evinces a want of taste, with much coarseness of feeling. Fullness over the eyes indicates excellent practical judgment of matters and things appertaining to property, science, and nature in general; while narrow, straight eyebrows indicate poor practical judgment of matter, things, their qualities, relations, and uses. Fullness from the root of the nose upward indicates great practical talent, love of knowledge, desire to see, and ability to say and do the right thing at the right time, and in the best way, together with sprightliness of mind; while a hollow in the middle of the forehead indicates want of memory, and inability to show off to advantage. A bold, high forehead indicates strong reasoning capabilities, while a retiring forehead indicates less soundness, but more availability of talent. And thus of other cerebral developments.

Phrenology teaches that every faculty, when active, moves head and body in the direction of the acting organ. Thus, intellect, in the fore part of the head, moves it directly forward, and produces a forward hanging motion of the head. Hence, intellectual men never carry their heads backward and upward, but always forward; and **logical** speakers move their heads in a straight line, usually forward, toward their audience; while vain speakers hold their heads backward. Hence it is a poor sign to stand so straight as to lean backward, for it shows that the brain is in the wrong place — more in the animal than intellectual region. Perceptive intellect, when active, throws out the chin and lower portions of the face; while reflective **intellect causes**

No. 139.—WASHINGTON IRVING.

the upper portion of the forehead to hang forward, and **draws in the** chin, as in Franklin, Webster, and other great thinkers. A coxcomb once asking a philosopher, "What makes you hang your head down so? why don't you hold it up as I do?" was answered: "Look at that field of wheat! The heads that are well filled bend downward, but those that stand up straight are empty." Benevolence throws the head and body slightly forward, leaning toward the object which excites its sympathy; while Veneration causes a low bow, which, the world over, is a token of respect; yet, when Veneration is **exercised** toward the Deity, as in devout prayer, it throws the head UPWARD; and, as we use intellect at the same time, the head is generally directed forward.

He who meets you with a long, low bow thinks more of you than of himself; but he who greets you with a short, quick bow — who makes half a bow forward, but a bow and a half backward — thinks one of you, and one and a half of himself. Ideality throws the head slightly forward and to one side, as in Irving, a man as gifted in taste and imagination as any other writer; and, in his portraits, his finger rests upon this faculty, while Sterne's finger rests upon Mirthfulness. Very firm men stand straight up and down, inclining not a hair's breadth forward or backward, or to the right or left; hence the expression, "He is an up-and-down man." And this organ is located exactly on a line with the body. Self-esteem, located in the back and upper portion of the head, throws the head and body upward and backward. Large feeling, pompous persons walk in a very dignified, majestic manner, throwing their heads in the direction of Self-esteem; while approbative persons throw their heads backward, but to one side. The difference between the natural language of these two organs is so slight that only the practical phrenologist can perfectly distinguish them.

The natural language of Money-loving, carries the head forward and to one side, as if in ardent pursuit of something, and ready to grasp it with outstretched arms; while Alimentiveness, situated lower, hugs itself down to the dainty dish with

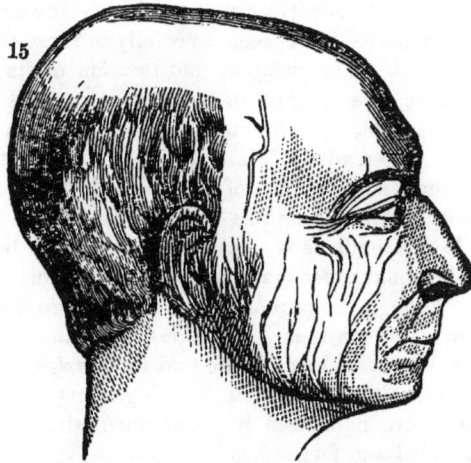

No. 140.—A CONCEITED SIMPLETON.

the greediness of an epicure, better seen than described. The shake of the head is the natural language of Combativeness, and means 'No, I resist you." Those who are combating earnestly shake the head more or less violently, according to the power of the combative feeling, but always shake it slightly inclining *backward;* while

Destructiveness, inclining forward, causes a shaking of the head slightly forward, and turning to one side. When a person who threatens you shakes his head violently, and holds it partially backward, and to one side, never fear—he is only barking; but whoever inclines his head to one side, and shakes it violently, will bite, whether possessed of two legs or four. Thus it is that each of the various postures assumed by individuals express the relative activity, present or permanent, of their respective faculties.

THE PHRENOLOGICAL FACULTIES, THEIR ANALYSIS AND CLASSI· FICATION.

But the highest, most conclusive evidence that Phrenology is true, is: Whatever is true bears indisputable evidence of its divine origin in its infinite perfection; while whatever is human is imperfect. If, therefore, Phrenology is true, every part and parcel of it will be perfection itself—in its facts, its philosophies, its teachings. And that proposed analysis of the phrenological faculties to which we now proceed will so expound its internal workings as to show whether it is or is not thus perfect or imperfect—true or false.

Its perfection is seen especially in these three aspects:—

First. In its grouping and location of its organs. Throughout all nature, the *location* of every organ serves to facilitate its function. Thus, foot, eye, heart, each bone and organ, can fulfill its office better placed where it is, than if placed anywhere else. Then if Phrenology is true, each of the phrenological organs will be so located, both absolutely and as regards the others, that their positions shall aid the ends they subserve. And their being thus placed furnishes additional proof that Phrenology is divine.

Though the phrenological organs were discovered, some in one century and continent, and others in another, yet on casting the analytical eye over them all, we find them *self*-classified by their topographical position in the head. Beginning at the lowest posterior organs they are numbered in accordance with their *topographical position* upward and forward.

And what is more, all those organs are in groups whose faculties perform analogous functions. Thus, all the social affections are grouped in one portion of the head—the back and lower; and their position is beneath and below all, just as their function is basilar, yet comparatively unseen. Neither do these organs obtrude themselves on our vision; nor do we stand on the corners of the streets to proclaim how much we love husband, wife, children, or friends. So the animal organs are placed at the top of the spinal column and base of the brain, or just where the nerves from the various portions of the body ramify on the brain. Now the office of these organs is to carry forward the various bodily wants. This nature fulfills, by placing them right at the head of those nerves which enable them to communicate with the body in the most perfect manner possible.

So the organs in the top of the head, being highest of all, fulfill the most exalted functions of all. By a law of structure, as we rise from the sole of the foot to the crown of the head, at every inch of our ascending progress we meet with functions more and still more important as their organs are located higher up. Feet, located lowest of all, perform the menial services of all; while the organs in the lower part of the body proper, higher in position, are also higher in function; for whereas we can live without feet, convenient though they be, yet we cannot live long without the visceral organs. Yet longer and better without these than without heart or lungs, which, located highest of all in the body proper, fulfill its most important functions, their suspension causing simultaneous death. But even these perform functions less elevated than the head, which, located highest of all, fulfills the crowning function of all — MIND; that for which the entire body, as well as universal nature, was created. And we might therefore infer that the various parts of this brain would fulfill functions more important, according to their position upward from the base to the top. And so it is. For while the animal and social organs are to man what foundations are to houses — absolutely indispensable — yet that there is a higher quality or grade to man's moral faculties than to his animal, to those which ally him to angels and to God than to matter, to immortality than mortality, is but the common sentiment of mankind. Is not the good man higher in the human scale than those who have only powerful animal functions? Are not those great intellectually greater than those great animally? The talented above the rich? Reason above Acquisitiveness? Does not the philosophy involved in this position of these various organs, both absolutely and as regards each other, evince a divine hand in its construction?

Secondly. Equally philosophical and perfect is the analysis of the phrenological faculties, considered both in reference to man's necessary life-requisitions, and as regards universal nature. Man having a material department to his nature, must needs be linked to matter, and possessed of all its properties. He is so. Then might we not expect some department of his nature to inter-relate him to each property of matter? These phrenological faculties furnish that relation. It so is that each phrenological faculty is adapted and adapts man to some great element in matter and arrangement in nature, and also to some special want or requisition of his being. Thus Appetite relates him to his need of food, and to that department of nature which supplies this food, or to her dietetic productions. Causality adapts him to nature's arrangements of cause and effect; Comparison, to her classifications; Form, to her configurations; Ideality, to the beautiful; and in like manner each of the other faculties adapts him to some institute of Nature. And to point out this adaptation furnishes the finest explanation of the faculties to be found, as well as the strongest proof that "the hand that formed them is divine." That is, Parental Love is adapted, and adapts man to, the infantile and parental relations. Nature must needs provide for the rearing of every individual child; and this she effects by creating in all parents — vegetable, animal, human — the parental sentiment, or love of their *own* young, particularly as infants, thus specifying just what adult shall care for each particular child, and absolutely providing for the rearing of all. Hence, whatever involves the relations of parents to their children comes under this faculty; and its correct analysis unfolds whatever concerns parents and their children. So Constructiveness adapts man to his need of clothes, houses, and materials for creature comforts, and is adapted to nature's mechanical institutes. And each of the other phrenological organs has a like adaptation to some great fact or provision in the economy of things.

And what is more yet, each phrenological faculty is found to run throughout all animal, all vegetable life, and to be an inherent property of things — of nature, of matter. Thus, the phrenological faculty of Firmness expresses a principle which runs throughout every phase of nature, as seen in the stability of all her operations — the perpetual return of her seasons, the immutability of her laws, the stability of her mountains, the uniformity and reliability or firmness of all her operations. Time, too, expresses a natural institute. For it not only appertains to man and all his habits — the natural period of his life included — but all plants are timed, observe each its own times and seasons. Each seed, fruit, animal, everything has its time. Some things begin and end their lives, as it were, in a day — others a year;

while the cedars of Lebanon or California live through many centuries. But even they have their germination, adolescence, maturity, decline, death, and decay. Given fruits ripen each at its given season; and even flowers and vegetables, transplanted from a southern to a northern latitude, keep up their periodical function in spite of oppo‑ site seasons. Has not every rock, even, its age, that is, a time ele‑ ment? Periodicity appertains to the earth, and to every one of its productions and their functions, as well as to every star — indeed, is a universal institute of nature. So is Order. For are not eye, foot, heart, spine, always in their respective places? And so of bark, root, limb, fruit, every organ of every animal and vegetable. That is, method is quite as much an element of universal nature as of man. Color is equally universal. So is Form. And is not Conscientious‑ ness in nature's arrangement that, all her laws obeyed, reward — vio‑ lated, punish? A tree injured inflicts punishment by withholding its fruit. And every wrong done to man, animal, or thing becomes its own avenger, while every right embodies its own reward, showing that the entity we call Conscientiousness is a universal institute, not of man alone, but of every phase of life and function of Nature. And so of all the other faculties.

Thirdly. Phrenology teaches the true *philosophy of life.* It unfolds the *original* constitution of man. That constitution was created just as perfect as its divine Author could render it. And in pointing out the original constitution of humanity, Phrenology shows who departs therefrom, and wherein. That is, by giving a *beau ideal* of human perfection, it teaches one and all, individuals and communities, wherein and how far they conform to, and depart from, this perfect human type, and thereby becomes the great reformer. And as far as individuals and communities live in accordance with its requisitions, they live *perfect* lives. That is, each of its faculties has a normal action which fulfilled is perfection, and also an abnormal, which is imperfection. And in teaching us both their normal and abnormal, it thereby teaches us just how to live, even in details; and thereby settles all questions in morals, in ethics, in all transactions between man and man, in every possible phase and aspect of life, down to its minutest details and requisitions, thereby becoming the great law‑ giver of humanity.

But to follow out these grand first principles would unduly enlarge our volume. Having stated them, the reader, curious to follow them up, will find in O. S. Fowler's "Phrenological Journal," and in his works on Phrenology, these and kindred ideas amplified. Meanwhile, to proceed with the phrenological organs, their groups, and individual functions.

THE SOCIAL GROUP, OR FAMILY AFFECTIONS.

These occupy the back and lower portion of the head, causing it to project behind the ears, and create most of the family affections and virtues.

7. — Are preëminently attached to family and home, and enjoy them more than any of the other pleasures of life; love companion and children with passionate fondness, and will do and sacrifice anything for them; and must have a home and home joys, and pine without them.

6. — Love family, home, country, and the fireside relations devotedly, and regard family as the centre of most of life's pleasures or pains; are eminently social and companionable, and strive to make home pleasant and family happy; and sacrifice often and much on the domestic altar.

5. — Love and enjoy the domestic relations well, but not as life's highest good; and seek other things and pleasures first, though home pleasures much.

4. — Have fair, average, commonplace family ties, and do much but not over much, for companion, children, and friends.

3. — Are rather indifferent in and to the family, and take a little, though no great pleasure in them; and need to cultivate the domestic virtues.

2. — Care little for home, its inmates, or pleasures, and are barren of its virtues.

1. — Have scarcely any social ties, and they weak.

1. AMATIVENESS.

THE CREATOR. — Sexuality; gender; the love element; that which attracts the opposite sex, and is attracted to it, admires and awakens admiration, creates and endows offspring, desires to love, be loved, and marry; the conjugal instinct and talent; gallantry; ladyism; masculinity in man, and womanliness in woman; passion. Adapted to Nature's male, female, sexual blending, affiliating, and creative ordinances.

Everything in nature is SEXED — male or female. And this sexual institute embodies those means employed by the Author of all life for its inception — for the perpetuity and multiplication of all forms of life. It creates in each sex admiration and love of the other; renders woman winning, persuasive, urbane, affectionate, loving, and lovely, and develops all the feminine charms and graces; makes man noble in feeling and bearing; elevated in aspiration; gallant, tender, and bland in manner; affectionate toward woman; highly susceptible to female charms; and clothes him with that dignity, power, and persuasiveness which accompanies the masculine.

VERY LARGE.   SMALL.

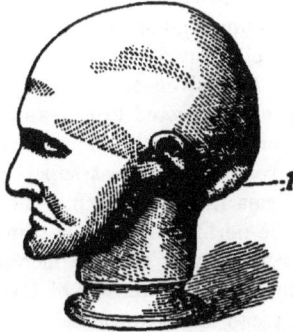 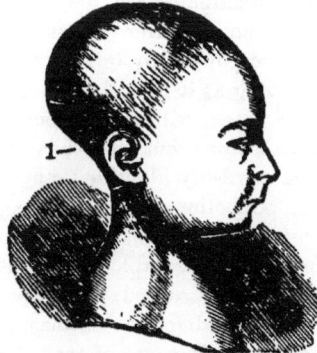

No. 141.—AARON BURR.   No. 142.—INFANT.

Perverted, it occasions grossness and vulgarity in expression and action; licentiousness in all its forms; a feverish state of mind; depraves all the other propensities; treats the other sex merely as a minister to passion — now caressing, and now abusing; and renders the love-feeling every way gross and animal.

VERY LARGE. — Are admirably sexed, or well-nigh perfect as a male or female; literally idolize the opposite sex; love almost to insanity; treat them with the utmost consideration; cherish for them the most exalted feelings of regard and esteem, as if they were superior beings; have the instincts and true spirit and tone of the male or female in a preëminent degree; must love and be beloved; love with inexpressible tenderness; are sure to elicit a return of love; are intuitively winning, attractive to, and attracted by, the other sex, in behavior, in conversation, in all they say and do; almost worship parents, brothers, or sisters, and children of the opposite sex; with organic quality 6, and the other social organs large, have the conjugal intuition in a preëminent degree; assimilate and conform to those loved, and become perfectly united; and with Conjugality large, manifest the most clinging fondness and utmost devotion, and are made or unmade for life by the state of the affections. For other combinations, see Large.

LARGE. — Are well sexed, or very much of a man or woman; that is, have the form, carriage, spirit, manners, and mind of the true male or female in a high degree; are eminently loving and lovely, or full of love, and with Conjugality large, of the real conjugal sentiment and intuition; strongly attract, and are strongly attracted by, the opposite sex; admire and love their beauty and excellences; easily win their affectionate regards, and enkindle their love; have many warm friends and admirers among them; love young and most intensely, and are powerfully influenced by the love element for good or evil, according as it is well or ill placed; with Adhesiveness and Conjugality large, will mingle pure friendship with devoted love; cannot flourish alone, but must have a matrimonial mate, with whom to become perfectly identified, and whom to invest with almost superhuman perfections, by magnifying their charms and overlooking their defects; in the sunshine of whose love to be perfectly happy, but proportionally miserable without it; with large Ideality and the mental temperament added, will experience a fervor and intensity of love, amounting almost to ecstasy or romance; can marry those only who combine refinement of manners with correspondingly strong attachments; with Parental Love and Benevolence also large, are eminently qualified to enjoy the domestic relations, and be happy in home, as well as to render home happy; with Inhabitiveness also large, will set a high value on house and place; long to return home when absent, and consider family and children as the greatest of life's treasures; with large Conscientiousness added, will keep the marriage relations inviolate, and regard unfaithfulness as the greatest of sins; with Combativeness large, will defend the object of love with great spirit, and resent powerfully any indignity offered them; with Alimentiveness large, will enjoy eating with loved one and family dearly; with Approbativeness large, cannot endure to be blamed by those beloved; with Cautiousness and Secretiveness large, will express love guardedly, and much less than is experienced; but with Secretiveness small, will show in every look and action the full, unveiled feelings of the soul; with Firmness, Self-esteem, and Conjugality large, will sustain interrupted love with fortitude, yet suffer much damage of mind and health therefrom; but with Self-esteem moderate, will feel crushed and broken down by disappointment; with the moral faculties predominant, can love those only whose moral tone is pure and elevated; with predominant Ideality, and only average intellectual faculties, will prefer those who are showy and gay to those who are sensible, yet less beautiful; but with Ideality less than the intellectual and moral organs, will prefer those who are substantial and valuable rather than showy; with Mirthfulness, Time, and Tune, will love dancing, lively company, etc.: p. 57.

FULL.— Possess quite strong susceptibilities of love for a congenial spirit; are capable of much purity, intensity, and cordiality of love, if its object is about right; with Adhesiveness and Benevolence large, will be kind and affectionate in the family; with Secretiveness large, will manifest less love than is felt, and show little in promiscuous society; with a highly susceptible temperament, will experience great intensity of love, and evince a good degree of masculine or feminine excellence, etc.: p. 59.

AVERAGE.— Are capable of fair conjugal attachments, and calculated to feel and exhibit a good degree of love, provided it is properly placed and fully called out, but not otherwise; experience a greater or less degree of love in proportion to its activity; as a man, are quite attached to mother, daughters, and sisters, and fond of female society, and endowed with a fair share of the masculine element, yet not remarkable for its perfection; as a woman, quite winning and attractive, yet not particularly susceptible to love; as a daughter, fond of father and brothers, and desirous of the society of men, yet not especially so; and capable of a fair share of conjugal devotedness under favorable circumstances; combined with an ardent temperament, and large Adhesiveness and Ideality, have a pure and platonic cast of love, yet cannot assimilate with a coarse temperament, or a dissimilar phrenology; are refined and faithful, yet have more friendship than passion; can love those only who are just to the liking; with Cautiousness and Secretiveness large, will express less love than is felt, and that equivocally, and by piecemeal, nor then till the loved one is fully committed; with Cautiousness, Approbativeness, and Veneration large, and Self-esteem small, are diffident in promiscuous society, yet enjoy the company of a select few of the opposite sex; with Adhesiveness, Benevolence, and Conscientiousness large, and Self-esteem small, are kind and affectionate in the family, yet not particularly fond of caressing or being caressed; and do much to make family happy, yet will manifest no great fondness and tenderness; with Order, Approbativeness, and Ideality large, seek in a companion personal neatness and polish of manners; with full intellectual and moral faculties, base their conjugal attachments in the higher qualities of the affections, rather than their personal attractiveness or strength of passion; but with a commonplace temperament, and not so full moral and intellectual faculties, are indifferent toward the opposite sex, and rather cool toward them in manners and conversation; neither attract nor are attracted much, are rather tame in love and marriage, and can live tolerably comfortable without loving or being beloved, etc.: p. 56.

MODERATE. — Are rather deficient, though not palpably so, in the love element, and averse to the other sex; and love their mental excellences more than personal charms; show little desire to caress or be caressed; and find it difficult to sympathize with a conjugal partner, unless the natural harmony between both is well-nigh perfect; care less for marriage, and can live unmarried without inconvenience; with Conjugality large, can love but once, and should marry the first love, because the love-principle will not be sufficiently strong to overcome the difficulties incident to its transfer, or the want of congeniality, and find more pleasure in other things than in the matrimonial relations; with an excitable temperament, will experience greater warmth and ardor than depth and uniformity of love; with Ideality large and organic quality 6, are fastidious and over-modest, and terribly shocked by allusions to love; pronounce love a silly farce, only fit for crack-brained poets; with Approbativeness large, will soon become alienated by rebukes and fault-finding; with Adhesiveness and the moral and intellectual faculties large, can become strongly attached to those who are highly moral and intellectual, yet experience no affinity for any other, and to be happy in marriage, must base it in the higher faculties: p. 59.

SMALL. — Dislike the opposite sex, and distrust and refuse to assimilate with them; feel little sexual love, or desire to marry; are cold, coy, distant, and reserved toward the other sex; experience but little of the beautifying and elevating influence of love, and should not marry, because incapable of appreciating its relations, and making a companion happy: p. 59.

VERY SMALL. — Are passively continent, and almost destitute of love: p. 60.

To CULTIVATE. — Mingle much in the society of the other sex; observe and appreciate their excellences, and overlook their faults; be as gallant, as gentlemanly or lady-like, as inviting, as prepossessing, as lively and entertaining in their society as you know how to be, and always on the alert to please them; say as many complimentary and pretty, and as few disagreeable things as possible; that is, try to cultivate and play the agreeable; if not married, contemplate its advantages and pleasures, and be preparing to enjoy them; if married, get up a second and an improved edition of courtship; reënamor both yourself and conjugal partner, by becoming just as courteous, loving, and lovely as possible; luxuriate in the company and conver-sation of those well sexed, and imbibe their inspiriting influence; be less fastidious, and more free and communicative; establish a warm, cordial intimacy and friendship for them, and feast yourself on their masculine or feminine excellences; if not married, marry, and cultivate the feelings, as well as live the life of a right and a hearty sexuality.

To RESTRAIN. — Simply *direct* this love element more to the mental, and less to the personal qualities of the other sex; admire and love them more for their minds than bodies, more for their moral purity and conversational powers than as instruments of passion; seek the society of the virtuous and good, but avoid that of the vulgar; should mingle in their society to derive moral elevation and inspiration therefrom, and be made better, not to feed the fires of passion; and yield to their moulding influences for good; should be content to commune with their *spirits*, should sanctify and elevate the cast and tone of love, and banish its baser forms; especially should lead a right *physiological* life — avoid tea and meats, and abstain wholly from coffee, tobacco, and all forms and degrees of alcoholic drinks, wines and beer included; exercise much in the open air; abstain wholly from carnal indulgence; work off your vital force on other functions as a relief of this; bathe daily; eat sparingly; study and commune with nature; cultivate the pure, intellectual, and moral as the best means of rising above the passional; and put yourself on a high human plane throughout. Remember these two things — first, that you require its purification, elevation, and right direction rather than restraint, because it is more perverted than excessive — it cannot be too great if rightly exercised — and secondly, that the inflamed state of the body irritates and perverts this passion, of which a cooling regimen is a specific antidote : p. 218.

2. CONJUGALITY.

FIDELITY. — Constancy in love, monogamy; union for life; first love; the pairing instinct; attachment to one conjugal partner; duality and exclusiveness of love.

Perverted action — a broken heart; jealousy; envy toward love rivals. Located between Amativeness and Adhesiveness, and adapted to parents living with and educating all their own children together in the same family. Some birds, such as doves, eagles, geese, robins, etc., pair, and remain true to their connubial attachment; while hens, turkeys, sheep, horses, and neat cattle associate promiscuously, which shows this to be a faculty distinct from Amativeness and Adhesiveness.

VERY LARGE. — Will select some ONE of the opposite sex as the sole object of love; concentrate the whole soul on this single *one* beloved, magnifying excellences and overlooking faults; long to be always with that one; are exclusive, and require a like exclusiveness; are true and faithful in wedlock, if married in spirit; possess the element of conjugal union, and flowing together of soul, in the highest degree, and with Continuity 6, become broken-hearted when disappointed, and comparatively worthless; seek death rather than life; regard this union as the gem of life, and its loss as worse than death; and should manifest the utmost care to bestow itself only where it can be reciprocated for life.

LARGE. — Seek one, and but one, sexual mate; experience keen disappointment when love is interrupted; are restless until the affections are anchored; are perfectly satisfied with the society of that one; and should exert every faculty to win the heart and hand of the one beloved; nor allow anything to alienate the affections.

FULL. — Can love cordially, yet are capable of changing the object, especially if Continuity is moderate; will love for life, provided circumstances are favorable, yet will not bear everything from a lover or companion, and if one love is interrupted can form another.

AVERAGE.—Are disposed to love but one for life, yet capable of changing their object, and, with Secretiveness and Approbativeness large, are capable of coquetry, especially if Amativeness is large, and Adhesiveness only full. Such should cultivate this faculty, by not allowing their other faculties to break their first love.

MODERATE. — Are somewhat disposed to love only one; yet allow other stronger faculties to interrupt that love, and, with Amativeness large, can form one attachment after another with comparative ease, yet are not true as a lover, nor faithful to the connubial union.

SMALL. — Have but little conjugal love, and seek the promiscuous society, and affection of the opposite sex, rather than a single partner for life.

VERY SMALL. — Manifest none of this faculty, and experience little.

To CULTIVATE. — Never allow new faces to awaken new loves, cut cling to the first one, and cherish its associations and reminis- cences; do not allow the affections to wander, but be much in the com- pany of the one already beloved, and both open your heart to love the charms, and keep up those thousand little attentions calculated to revive and perpetuate conjugal love : p. 230.

To RESTRAIN. — Seal up and bury the volume of your first affec- tion. Another will then take its place ; try to appreciate the excel- lences of others than the first love, remembering that " there are as good fish in the sea as ever were caught ; " if a first love dies or is blighted, by no means allow yourself to pore over the bereavement, but transfer affection just as soon as a suitable object can be found, and be industrious in finding one, by making yourself just as accept- able and charming as possible. Above all, do not allow a pining, sad feeling to crush you, nor allow hatred toward the other sex : p. 230.

3. PARENTAL LOVE.

THE NURSE. — Attachment to own offspring; love of chil- dren, the young, and pets; playfulness with them.

Adapted to that infantile condition in which man enters the world, and to children's need of parental care and education This faculty renders children the richest treasure of their parents; casts into the shade all the toil and expense they cause; and lacerates them with bitter pangs when death or dis- tance tears them asunder. It is much larger in woman than in man ; and nature requires mothers to take the principal care of 'nfants. Perverted, it spoils children by excessive indulgence, pampering, and humoring.

VERY LARGE. — Love their own children with the greatest possible intensity and pathos; almost idolize their own children, grieve

VERY LARGE.   DEFICIENT.

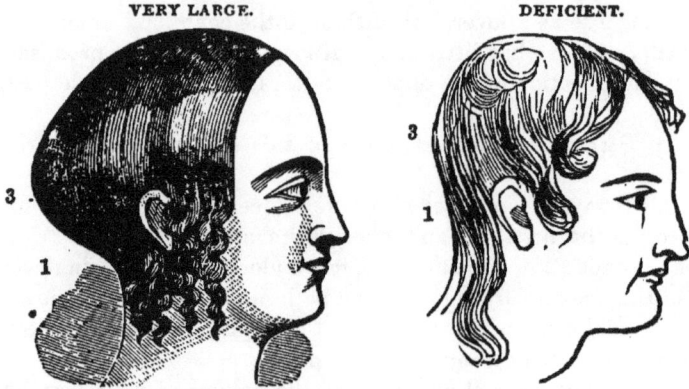

No. 143. — THE DEVOTED MOTHER.   No. 144. — THE UNMOTHERLY.

immeasurably over their loss, and with large Continuity, refuse to be comforted; with very large Benevolence, and only moderate Destructiveness, cannot bear to see them punished, and with only moderate Causality, are liable to spoil them by over-indulgence; with large Approbativeness added, indulge parental vanity and conceit; with large Cautiousness and disordered nerves, caution them continually, and feel a world of groundless apprehensions about them; with Acquisitiveness moderate, make them many presents, and lavish money upon them; but with large Acquisitiveness, lay up fortunes for them; with large moral and intellectual organs, are indulgent, yet love them too well to spoil them; and do their utmost to cultivate their higher faculties, etc.: p. 63.

LARGE. — Love their own children devotedly; value them above all price; cheerfully endure toil and watching for their sake; forbear with their faults; win their love; delight to play with them, and cheerfully sacrifice to promote their interests; with Continuity large, mourn long and incessantly over their loss; with Combativeness, Destructiveness, and Self-esteem large, are kind, yet insist on being obeyed; with Self-esteem and Destructiveness moderate, are familiar with, and liable to be ruled by them; with Firmness only average, manage them with an uneven hand; with Cautiousness large, suffer extreme anxiety if they are sick or in danger; with large moral and intellectual organs, and less Combativeness and Destructiveness, govern them more by moral suasion than physical force — by reason than fear; are neither too strict nor over-indulgent; with Approbativeness large, value their moral character as of the utmost importance; with Veneration and Conscientiousness large, are particularly interested in their moral improvement; with large excitability, Combativeness, and Destructiveness, and only average Firmness, will be, by turns, too indulgent, then over-provoked — pet them one minute, but punish them the next; with larger Approbativeness and Ideality than intellect, educate them more for show than usefulness — more fashionably than substantially — and dress them off in the extreme of fashion; with a large and active brain, large moral and intellectual faculties, and Firmness, and only full Combativeness, Destructiveness, and Self-esteem, are well calculated to teach and manage the young. It renders farmers fond of stock, dogs, etc., and women of birds, lap-dogs, etc. ; girls of dolls, and boys of being among horses and cattle; and love the young, weak, and petite : p. 62.

FULL. — Love their own children well, yet not passionately — do much for them, yet not more than necessary — and with large Combativeness, Destructiveness, and Self-esteem, are too severe, and make too little allowance for their faults ; but with Benevolence, Adhesiveness, and Conscientiousness large, do and sacrifice much to supply their wants and render them happy. Its character, however, will be mainly determined by its combinations : p. 63.

AVERAGE. — Love their own children tolerably well, yet care but little for those of others ; with large Adhesiveness and Benevolence, like them better as they grow older, yet do and care little for infants ; are not duly tender to them, or forbearing toward their faults, and should cultivate parental fondness, especially if Combativeness, Destructiveness and Self-esteem are large: p. 61.

MODERATE. — Are not fond enough of children ; will not bear much from them ; fail to please or take good care of them, particularly of infants ; cannot endure to hear them cry, or make a noise, or disturb things ; and with an excitable temperament, and large Combativeness, are liable to punish them for trifling offenses, find much fault with them, and be sometimes cruel ; yet, with Benevolence and Adhesiveness large, do what is necessary for their comfort : p. 64.

SMALL. — Care little for their own children, and still less for those of others ; and with Combativeness and Destructiveness large, are liable to treat them unkindly and harshly, and are utterly unqualified to have charge of them : p. 64.

VERY SMALL. — Have little or no parental love or regard for children, but conduct toward them as the other faculties dictate : p. 64.

TO CULTIVATE. — Play with and make much of children ; try to appreciate their loveliness and innocence, and be patient, tender, and indulgent toward them ; and if destitute of own children, adopt some, or provide something to pet and fondle : 221.

TO RESTRAIN. — Should set judgment over against affection ; rear them intellectually ; feel less anxiety about them, and if a child dies, by all means turn from that loss by seeking some powerful diversion, and a change of associations, removing clothes and all remembrances, and not talk or think about them

4. FRIENDSHIP.

THE CONFIDANT. — Sociability ; love of society ; desire to congregate, associate, visit, make and entertain friends, etc.

When perverted it forms attachments for the unworthy, and seeks bad company. It is adapted to man's requisition for concert of action, copartnership, combination, and community of feeling and interest, and is a leading element in his social relations.

VERY LARGE. — Love friends with the utmost tenderness and intensity, and will sacrifice almost anything for their sake; with Amativeness large, are susceptible of the highest order of conjugal love, yet base that love primarily in friendship ; with Combativeness and Destructiveness large, defend friends with great spirit, and resent and retaliate their injuries; with Self-esteem moderate, take character from associates ; with Acquisitiveness moderate, allow friends free use of purse; but with Acquisitiveness large, will do more than give ; with Benevolence and Approbativeness moderate, and Acquisitiveness only full, will spend money freely for social gratification ; with Self-esteem and Combativeness large, must be first or nothing ; but with only average Combativeness, Destructiveness, and Self-esteem, large Approbativeness, Benevolence, Conscientiousness, Ideality, and reasoning organs, make many friends, and few enemies; are amiable and universally beloved ; with large Eventuality and Language, recount with vivid emotions, by-gone scenes of social cheer and friendly converse ; with large reasoning organs, give good advice to friends, and lay excellent plans for them ; with smaller Secretive ness and large moral organs, believe no ill of them, and dread the interruption of friendship as the greatest of calamities; willingly make any sacrifice required by friendship, and evince a perpetual flow of that commingling of soul, and desire to become one with others, which this faculty inspires: p. 65.

LARGE. — Are warm, cordial, and ardent in friendship; readily form acquaintances, and attract friendly regards in return; must have society of some kind; with Benevolence large, are hospitable and delight to entertain friends; with Alimentiveness large, love the social banquet, and set the best before friends; with Approbativeness large, prize their commendation, but are terribly cut by their rebukes; with the moral faculties large, seek the society of the moral and elevated, and can enjoy.the friendship of no others; with the intellectual faculties large, seek the society of the intelligent; with Language large, and Secretiveness small, talk freely in company; and with Mirthfulness and Ideality also large, are full of fun, and give a lively, jocose turn to conversation, yet are elevated and refined; with Self-esteem large, lead off in company, and give tone and character to others; but with Self-esteem small, receive character from friends, and with Imitation large, are liable to copy their faults as well as virtues; with Cautiousness, Secretiveness, and Approbativeness large, are apt to be jealous of regards bestowed upon others, and exclusive in the choice of friends — having a few select, rather than many commonplace; with large Causality and Comparison, love philosophical conversation, literary societies, etc., and are every way sociable and companionable: p. 65.

FULL. — Make a sociable, companionable, warm-hearted friend, sacrifice much on the altar of friendship, yet offer up friendship on the altar of the stronger passions; with large or very large Combativeness, Destructiveness, Self-esteem, Approbativeness, and Acquisitiveness, serve self first, and friends afterward; form attachments, and break them when they conflict with the stronger faculties; with large Secretiveness and moderate Conscientiousness, are double-faced, and profess more friendship than is felt; with Benevolence large, cheerfully aid friends, yet more from sympathy than affection; have a few warm friends, yet only few, but perhaps many speaking acquaintances; and with the higher faculties generally large, are true, good friends, yet by no means enthusiastic. Many of the combinations under Adhesiveness large, apply to it when full, allowance being made for its diminished power: p. 66.

AVERAGE. — Are capable of quite strong friendships, yet their character is determined by the larger faculties; enjoy present friends, yet sustain their absence; with large Acquisitiveness, place business before friends, and sacrifice them whenever they conflict with money making; with Benevolence large, are more kind than affectionate, relish friends, yet sacrifice no great for their sake; with Amativeness large, love the person of the other sex more than their minds, and experience less conjugal love than animal passion; with Approbativeness large, break friendship when ridiculed or rebuked, and with Secretiveness large, and Conscientiousness only average, cannot be trusted in friendships : p. 64.

MODERATE. — Love society somewhat, and form a few, but only few, attachments, and these only partial; may have many speaking acquaintances, but make few intimate friends; with large Combativeness and Destructiveness, are easily offended with friends, and seldom retain them long; with large Benevolence, bestow services, and with moderate Acquisitiveness, money more readily than affection; but with the selfish faculties strong, take care of self first, and make friendship subservient to interest: p. 67.

SMALL. — Think and care little for friends; dislike copartnership, are cold-hearted, unsocial, and selfish; take little delight in company; prefer to be alone; have few friends, and with large selfish faculties, many enemies, and manifest too little of this faculty to exert a perceptible influence upon character: p. 67.

VERY SMALL. — Are perfect strangers to friendship: p. 67.

TO CULTIVATE. — Go more into society; associate freely with those around you; open your heart; be less exclusive and distant; keep your room less, but go more to parties, and strive to be as companionable and familiar as you well can; nor refuse to affiliate with those not exactly to your liking, but like what you can, and overlook faults : 226.

TO RESTRAIN. — Go abroad less, and be more select in choosing friends; besides guarding yourself against those persuasions and influences friends are apt to exercise over you, and trust friends less, as well as properly direct friendship by intellect : 227.

5. INHABITIVENESS.

LARGE.

THE PATRIOT. — Love of domicil, and country; of home, house, and the place where one lives and has lived.

The HOME feeling; love of HOUSE, the PLACE where one was born or has lived, and of home associations. Adapted to man's need of an abiding place, in which to exercise the family feelings; patriotism. Perversion — homesickness when away from home, and needless pining after it.

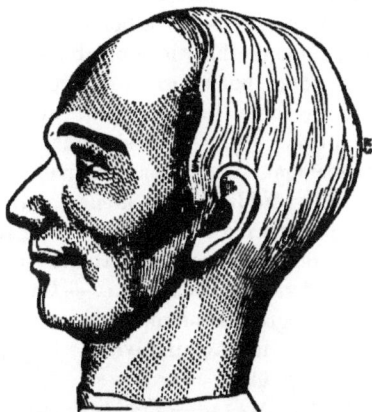

No. 145. — CLAY, THE PATRIOT.

VERY LARGE. — Are liable to homesickness when away from home, especially for the first time, and the more so if Parental Love and Adhesiveness are large; will suffer almost any inconvenience, and forego bright prospects rather than leave home; and remain in an inferior house or place of business rather than change. For combinations, see Inhabitiveness large: p. 68.

LARGE. — Have a strong desire to locate young, and have a home or room exclusively; leave home with great reluctance, and return with extreme delight; soon become attached to house, sleeping-room, garden, fields, furniture, etc., and highly prize domestic associations; are not satisfied without a place on which to expend this home instinct; with Parental Love, Adhesiveness, Individuality, and Locality large, will love to travel, yet be too fond of home to stay away long at a time; may be a cosmopolite in early life, and love to see the world, but will afterward settle down; with Approbativeness and Combativeness large, will defend national honor, praise own country, government, etc., and defend both country and fireside with great spirit; with Ideality large, will beautify home; with Friendship large, will delight to see friends at home rather than abroad; with Alimentiveness large, will better enjoy food at home than elsewhere, etc.. p. 68.

FULL. — Prefer to live in one place, yet willingly change it when interest or the other faculties require; and with large Parental Love, Adhesiveness, and Amativeness, will think more of family and friends than of the domicile : p. 69.

AVERAGE. — Love home tolerably well, yet with no great fervor, and change the place of abode as the other faculties may dictate; take some, but no great interest in house or place, as such, or pleasure in their improvement, and are satisfied with ordinary home comforts; with Acquisitiveness large, spend reluctantly for its improvement; with Constructiveness moderate, take little pleasure in building additions to home; with Individuality and Locality large, love travelling more than staying in one place, and are satisfied with inferior home accommodations : p. 63.

MODERATE OR SMALL. — Care little for home; leave it without much regret; contemplate it with little delight; take little pains with it; and with Acquisitiveness large, spend reluctantly for its improvement : p. 69.

VERY SMALL. — Feel little, and show less, love of domicile, as such.

TO CULTIVATE. — Stay more at home, and cherish a love of it and its associations and joys, and also love of country : 232.

TO RESTRAIN. — Go from home, and banish that feeling of home-sickness experienced away from home, by diversions : 233.

6. CONTINUITY.

THE FINISHER. — Consecutiveness; connectedness; poring over one thing till it is done; prolixity; unity.

Dwelling patiently upon one thing till it is done; consecutiveness and connectedness of thought and feeling. Adapted to man's need of doing one thing at a time. Perversion — prolixity, repetition, and excessive amplification.

VERY LARGE. — Fix the mind upon objects slowly, yet cannot leave them unfinished; have great application, yet lack intensity or point; are tedious, prolix, and thorough in a few things, rather than an amateur in many : p. 70.

LARGE. — Give the whole mind to the one thing in hand till it is finished; complete at the time; keep up one common train of thought, or current of feeling, for a long time; are disconcerted if attention is directed to a second object, and cannot duly consider either; with Adhesiveness large, pore sadly over the loss of friends for months

CONTINUITY LARGE.

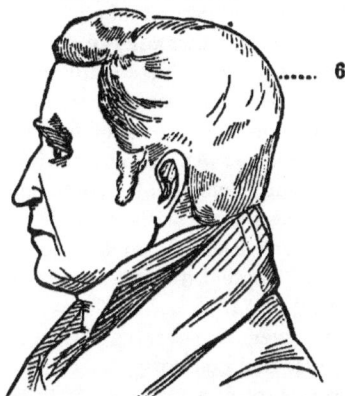

No. 146. — REV. DR. BUSH.

and years; with the moral faculties large, are uniform and consistent in religious exercises and character; with Combativeness and Destructiveness large, retain grudges and dislikes for a long time; with Ideality, Comparison, and Language large, amplify and sustain figures of speech; with the intellectual faculties large, con and pore over one subject of thought or study, and impart a unity and completeness to intellectual investigations; become thorough in whatever is commenced, and rather postpone until sure of completing : p. 70.

FULL. — Dwell continuously upon subjects, unless especially called to others; prefer to finish up matters in hand, yet can, though with difficulty, give attention to other things; with the business organs large, make final settlements; with the feelings strong, continue their action, yet are not monotonous, etc. : p. 71.

AVERAGE. — Can dwell upon one thing, or divert attention to several, as occasion requires; are not confused by interruption, yet prefer one thing at a time; with the intellectual organs large, are not smatterers, nor yet profound; with the mental temperament, are clear in style, and consecutive in idea, yet never tedious; with Comparison large, manufacture expressions and ideas consecutively and connectedly, and always to the point, yet never dwell unduly : p. 70.

MODERATE. — Love and indulge variety and change of thought, feeling, occupation, etc.; are not confused by them; rather lack application; with a good intellectual lobe and an active temperament, learn and do a little about a good many things, rather than much about any one thing; think clearly, and have unity and intensity of thought and feeling, yet lack connectedness; with large Language and small Secretiveness, talk easily, but not long at a time upon any one thing; do better on the spur of the moment than by previous preparation; and should cultivate consistency of character and fixedness of mind, by finishing all begun : p. 71.

SMALL. — With Activity great, commence many things, yet finish

few; crave novelty and variety; thrust many irons in the fire; lack
application; jump rapidly from premise to conclusion, and fail to
connect and carry out ideas; lack steadiness and consistency of character; may be brilliant, yet cannot be profound; humming-bird like,
fly rapidly from thing to thing, but do not stay long; have many good
thoughts, yet they are scattered; and talk on a great variety of subjects in a short time, but fail sadly in consecutiveness of feeling,
thought and action. An illustrative anecdote: An old and faithful
servant to a passionate, petulant master finally told him he could endure his testiness no longer, and must leave, though with extreme reluctance. "But," replied the master, "you know I am no sooner
angry than pleased again." "Aye, but," replied the servant, "you
are no sooner pleased than angry again: " p. 71.

VERY SMALL.—Are restless, and given to perpetual change;
with Activity great, are composed of gusts and counter-gusts of passion, and never one thing more than an instant at a time: p. 72.

TO CULTIVATE. — Dwell on, and pore over, till you complete the
thing in hand; make thorough work; and never allow your thoughts
to wander, or attention to be distracted, or indulge diversity or variety in anything: p. 284.

TO RESTRAIN. — Engage in what will compel you to attend to a
great many different things in quick succession. and break up that
prolix, long-winded monotony caused by its excess: p. 234.

SELFISH PROPENSITIES.

These provide for man's bodily wants; create our animal desires and instincts, and supply those wants which relate more
especially to his physical necessities.

VERY LARGE. — Experience great intensity of the animal impulses; enjoy personal existence and pleasures with the keenest relish; and with great excitability or a fevered state of body, are strongly predisposed to sensual gratifications and passional desires; yet if
properly directed, and sanctified by the higher faculties, have tremendous force of character and energy of mind.

LARGE. — Have strong animal desires; and that selfishness which
takes good care of number one; are strongly attached to this world
and its pleasures; and with activity great, use vigorous exertions to
accomplish worldly and personal ends; with the moral organs less

than the selfish, connected with bodily disease, are liable to their depraved and sensual manifestation ; but with the moral and intellectual large, and a healthy organization, have great force, energy, determination, and that efficiency which accomplishes wonders.

FULL. — Have a good share of energy and physical force, yet no
more than is necessary to cope with surrounding difficulties; and
with large moral and intellectual faculties, manifest more mental than
physical power.

AVERAGE. — Have a fair share of animal force, yet hardly enough
to grapple with life's troubles and wrongs; with large moral and intellectual faculties, have more goodness than efficiency, and enjoy
quiet more than conflict with men; and fail to manifest what goodness and talent are possessed.

MODERATE. — Rather lack efficiency ; yield to difficulties ; need
more fortitude and determination ; fail to assert and maintain rights ·
and with large moral organs, are good-hearted, moral, etc., yet border
on tameness.

SMALL. — Accomplish little; lack courage and force, and with
large intellectual organs, are talented, yet utterly fail to manifest that
talent; and with large moral organs, are so good as to be good for
nothing.

To CULTIVATE. — Keep a sharp eye on your own interests ; look
out well for number one; fend off imposition; harden up; don't be
so good ; and in general cultivate a burly, driving, self-caring, physical, worldly spirit ; especially increase the physical energies by observing the health laws, as this will reincrease these animal desires.

To RESTRAIN. — First and most, obviate all causes of physical
inflammation and false excitement; abstain from spirituous liquors,
wines, tobacco, mustards, spices, all heavy and rich foods ; eat lightly,
and of farinaceous rather than of flesh diet, for meat is directly
calculated to excite the animal passions; avoid temptation and incentives to anger and sensuality; especially associate only with the
good, never with those who are vulgar or vicious; but most of all,
cultivate the higher, purer moral faculties, and aspire to the high and
good ; also cultivate love of Nature's beauties and works, as the very
best means of restraining the animal passions.

7. VITATIVENESS.

THE DOCTOR. — Love and tenacity of life ; resistance to

disease and death ; dread of annihilation ; clinging tenaciously to existence, for its own sake ; toughness ; constitution.

VERY LARGE. — Shrink from death, and cling to life with desperation : struggle with the utmost determination against disease and death ; never give up to die till the very last, and then only by inches ; with Cautiousness very large, and Hope moderate, shudder at the very thought of dying, or being dead ; but with Hope large, expect to live against hope. The combinations are like those undei Large, allowance being made for the increase of this faculty. p.

LARGE. — Struggle resolutely through fits of sickness, and will not give up to die till absolutely compelled to do so. With large animal organs, cling to life on account of this world's gratifications; with large moral organs, to do good — to promote human happiness, etc.; with large social faculties, love life both for its own sake and to bless family; with very large Cautiousness, dread to change the present mode of existence, and with large and perverted Veneration and Conscientiousness, and small Hope, have an indescribable dread of entering upon an untried future state; but with Hope large, and a cultivated intellect, expect to exist hereafter, etc.

FULL. — Love life, and cling tenaciously to it, yet not extravagantly; are loth to die, and yield to disease and death, though reluctantly.

AVERAGE. — Enjoy life, and cling to it with a fair degree of earnestness, yet by no means with passionate fondness ; and with a given constitution and health, will die easier and sooner than with this organ large.

MODERATE OR SMALL. — Like to live, yet care no great about existence for its own sake ; with large animal or domestic organs, may wish to live on account of family, or business, or worldly pleasures, yet care less about it for its *own sake*, and yield it up with little reluctance or dread.

VERY SMALL. — Have no desire to live merely for the sake of living, but only to gratify other faculties.

TO CULTIVATE. — Think on the value of life, and plan things to be done and pleasure to be enjoyed worthy to live for : 236.

TO RESTRAIN. — Guard against a morbid love of life, and dread of death, but regard death as much as possible as a natural institution, and this life as the pupilage for a better state of being : 237.

8. COMBATIVENESS

THE DEFENDER. — Courage persistence; boldness; resist-ance; defense; self-protection; spirit; desire to encounter; love of opposition; defiance; determination; presence of mind; get-out-of-my-way; let-me-and-mine-alone. Adapted to man's requisition for over-coming obstacles, contending for rights, etc. Perversion — wrath; contrariety; fault-find-

No. 147. — VERY LARGE.

ing; contention; ill-nature; and fighting.

VERY LARGE. — Show always and everywhere the utmost hero-ism, boldness, and courage; can face the cannon's mouth coolly, and look death in the face without flinching; put forth remarkable efforts in order to carry measures; grapple right in with difficulties with a real relish, and dash through them as if mere trifles; love pioneer life and adventurous, even hazardous expeditions; shrink from no danger; are appalled by no hardships; prefer a rough and daring life — one of struggle and hair-breadth escapes — to a quiet, mo-notonous business; are determined never to be conquered, even by superior odds, but incline to do battle single-handed against an army; with Cautiousness only full, show more valor than discretion, are often fool-hardy, and always in hot water; with smaller Secretive-ness and Approbativeness, are most unamiable, if not hateful; with drinking habits and bad associates, have a most violent, ungoverna-ble temper; are desperate, most bitter, and hateful, and should never be provoked. For additional combinations see large, allowing for difference in size: p. 77.

LARGE. — Are bold, resolute, fearless, determined, disposed to grapple with and remove obstacles, and drive whatever is under-taken; love debate and opposition; are perfectly cool and intrepid; have great presence of mind in times of danger, and nerve to encoun-ter it; with large Parental Love, take the part of children; with large Inhabitiveness, defend country; with activity large and vitality mod-

erate, overdo perpetually, and should throw far less vim into efforts
with a powerful muscular system, put forth all the strength in lifting
working, and all kinds of manual labor; with great Vitativeness and
Destructiveness, defend life with desperation, and strike irresistible
blows; with large Acquisitiveness, maintain pecuniary rights; and
drive money-making plans; with large Approbativeness, resent in-
sults, and large Adhesiveness added, defend the character of friends;
with full or large Self-esteem, defend personal interests, take own
part with spirit, and repel all aggressions; with Self-esteem small,
and Benevolence and Friendship large, defend the interests of friends
more than of self; with large Conscientiousness, prosecute the right,
and oppose the wrong most spiritedly; with large intellectual organs,
impart vigor, power, and impressiveness to thoughts, expressions,
etc. ; with disordered nerves, are peevish, fretful, fault-finding,
irritable, dissatisfied, unreasonable, and fiery in anger, and should
first restore health, and then restrain this fault-finding disposition, by
remembering that the cause is internal, instead of what is fretted
at : p. 75.

FULL. — Evince those feelings described under large, yet in a less
degree, and as modified more by the larger organs; thus, with large
moral and intellectual faculties, show much more moral than physical
courage; maintain the right and oppose the wrong; yet, with Firmness large, in a decided rather than a combative spirit, etc.; p. 78.

AVERAGE. — Evince the combative spirit according to circumstances; when vigorously opposed, or when any of the other faculties
work in conjunction with Combativeness, show a good degree of the
opposing, energetic feeling; but when any of the other faculties, such
as large Cautiousness or Approbativeness, work against it, are irresolute, and even cowardly; with an active temperament, and disordered
nerves, especially if dyspeptic, have a quick, sharp, fiery temper, yet
lack power of anger; will fret and threaten, yet mean and do but little;
with a large brain, and large moral and intellectual organs, will put
forth fair intellectual and moral force when once thoroughly roused,
which will be but seldom; with large Approbativeness, and small Acquisitiveness, will defend character, but not pecuniary rights; with
large Cautiousness, may be courageous where danger is far off, yet
will run rather than fight; with smaller Cautiousness, will show some
resentment when imposed upon, but submit rather tamely to injuries;
with very large Parental Love, and only average Friendship, will resent any injuries offered to children with great spirit, yet not those
offered to friends, etc. : p. 75.

MODERATE. — Rather lack efficiency; with only fair muscles, **are** poor workers, and fail to put forth even the little strength possessed; with good moral and intellectual organs, possess talent and moral worth, yet are easily overcome by opposition or difficulty; should seek some quiet occupation, where business comes in of itself, because loth to intrude unbidden upon the attention of others; are too good to be energetic; with weak Acquisitiveness, allow virtual robbery without resentment; with large Cautiousness, are tame and pusillanimous; with large Approbativeness, cannot stand rebuke, but wilt under it; with moderate Self-esteem and Hope, are all " I can't, it's hard," etc., and will do but little in life: p. 78.

SMALL. — Are inert and inefficient; can accomplish little; never feel self-reliant or strong; and with large moral and intellectual organs, are too gentle and easily satisfied; with large Cautiousness, run to others for protection, and are always complaining of bad treatment: p. 79.

VERY SMALL. — Possess scarcely any energy, and manifest none: p. 79.

To CULTIVATE. — Encourage a bold, resistant, defiant, self-defending spirit; fend off imposition like a real hero; rather encourage than shrink from encounter; engage in debate, and the mental conflict of ideas and sentiments in politics, in religion, in whatever comes up, and take part in public meetings; takes sides in everything; say and try to feel, "None shall provoke me with impunity: " 239.

To RESTRAIN. — Do just the opposite of the preceding advice; whenever you find anger rising, turn on your heel; avoid debate, and say mildly and pleasantly whatever you have to say; bear with imposition rather than resent it; cultivate a turn-the-other-cheek spirit, never swear, or scold, or blow up anybody, and restrain temper and wrath in all their manifestations: 240.

9. DESTRUCTIVENESS.

THE EXTERMINATOR. — Executiveness; severity; sternness; harshness; love of tearing down, destroying, causing pain, teasing etc.; hardihood; endurance of pain; force.

Adapted to man's need of destroying whatever is prejudicial to his happiness; performing and enduring surgical operations; undergoing pain, etc. Perversion — violence; revenge malice; disposition to murder. etc.

VERY LARGE. — Experience the most powerful indignation, amounting, when thoroughly provoked, even to rage and violence ; and with large or very large Combativeness, act like a chafed lion, and

VERY LARGE.
16

APPETITE LARGE.

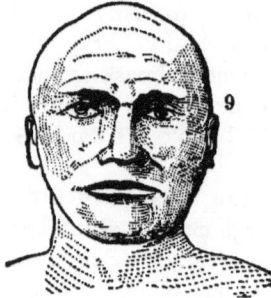

No. 148. — BLACK HAWK.   No. 149. — ROOT EATING INDIAN.

feel like rushing into the midst of perilous dangers ; tear up and destroy whatever is in the way ; are rough, harsh, and often morose in manner, and should cultivate pleasantness ; with large Combativeness, Firmness, Self-esteem, and Approbativeness moderate, are exceedingly repulsive, hating and hateful when angry, and much more provoked than occasion requires ; with large intellectuals, put forth tremendous mental energy ; and should offset this faculty by reason and moral feeling, and cultivate blandness instead of wrath : p. 83.

LARGE. — Impart that determination, energy, and force which remove or destroy whatever impedes progress ; with Firmness large, have that iron will which endures till the very last, in spite of everything, and will carry points anyhow ; with large Combativeness, impart a harsh, rough mode to expression and action, and a severity if not fierceness to all encounters ; with large Acquisitiveness and Conscientiousness, will have every cent due, though it cost two to get one, yet want no more, and retain grudges against those who have injured the pocket ; with large Approbativeness and Combativeness, feel determined hostility toward those who trifle with reputation or impeach character ; with large Self-esteem, against those who conflict with its interests, or retract from its merits ; with large Adhesiveness, when angry with friends, are angry forever ; with large Benevolence and Conscientiousness, employ a harsh mode of showing kindness ; with large Comparison and Language, heap very severe and galling epithets

upon enemies; with large Ideality, polish and refine expression of
anger, and put a keen edge upon sarcasms, yet they cut to the very
bone, etc. Such should avoid and turn from whatever provokes: p
82.

FULL. — Evince a fair degree of this faculty, yet its tone and direc
tion depend upon the larger organs; with large propensities, manifest much animal force; with large moral organs, evince moral determination and indignation; with large intellectual organs, possess
intellectual might and energy, and thus of its other combinations;
but with smaller Combativeness, are peaceful until thoroughly rouzed,
but then rather harsh and vindictive: if boys, attack only when sure
of victory, yet are then harsh; with smaller Self-esteem, exercise this
faculty more in behalf of others than of self; with large Cautiousness,
and moderate Combativeness, keep out of danger, broils, etc., till
compelled to engage in them, but then become desperate, etc.: p. 83.

AVERAGE. — Are like Full, only less so: p. 82.

MODERATE. — Evince but little harshness or severity, and shrink
from pain; with large Benevolence, are unable to witness suffering or
death, much less to cause them; possess but little force of mind or
executiveness of character, to drive through obstacles; with large
moral organs added, are more beloved than feared, manifest extreme
sympathy, amounting sometimes even to weakness, and secure ends
more by mild than severe measures; with moderate Combativeness
and Self-esteem, are irresolute, unable to stand ground, or take care
of self; fly to others for protection; can do little, and feel like trying
to do still less — fail to realize or put forth strength; and with large
Cautiousness added, see lions where there are none, and make mountains
tains of mole-hills; and with small Hope added, are literally good for
nothing; but with large Hope and Firmness, and full Self-esteem and
Combativeness, accomplish considerable, yet in a quiet way, and by
perseverance more than force — by siege rather than by storm — and
with large intellectual and moral faculties added, are good, though
not tame; exert a good influence, and that always healthful, and are
mourned more when dead than prized while living. The combinations
under this organ large, reversed, apply to it when moderate: p. 84.

SMALL. — With large moral faculties, possess too tender a soul to
enjoy our world as it is, or to endure hardships or bad treatment; can
neither endure nor cause suffering, anger being so little as to provoke
only ridicule, and need hardness and force: p. 82.

VERY SMALL. — Experience little, and manifest none of this faculty.
ulty.

To CULTIVATE. — Destroy anything and everything in your way. Killing weeds, blasting rocks, felling trees, using edge-tools, tearing up roots, plowing new ground, cultivating new farms, hunting, exercising indignation when wronged, and against public wrongs; espousing the cause of the oppressed; fighting public evils, such as intemperance and the like, are all calculated to cultivate and strengthen this faculty. Still, care should be taken to exercise it under the control of the higher faculties, and then no matter how great that exercise: 242.

To RESTRAIN. — Kill nothing; and offset Destructiveness by Benevolence; never indulge a rough, harsh spirit, but cultivate instead a mild and forgiving temper; never brood over injuries or indulge revengeful thoughts or desires, or aggravate yourself by thinking over wrongs; cultivate good manners; and when occasion requires you to reprove, do it in a bland, gentle manner rather than roughly; never tease, even children, or scourge animals, but be kind to both, and offset by Benevolence and the higher faculties: 243.

10. ALIMENTIVENESS.

No. 150. — VERY LARGE.

KING LOUIS XIV. A GREAT BANQUETER.

No. 151. — SMALL.

A POOR FEEDER.

THE FEEDER. — Appetite; the feeding instinct; relish for food; hunger. Adapted to man's need of food, and of an eating instinct. Perverted, it produces gormandizing and gluttony, and causes dyspepsia with all its evils.

VERY LARGE. — Often eat more than is requisite; enjoy food exceedingly well; and hence are liable to clog body and mind by overeating; should restrain appetite; will feel better by going without an occasional meal, and are liable to dyspepsia. This faculty is liable to take on a diseased action, and crave a much greater amount of food than nature requires, and hence is the great cause of dyspepsia. Its diseased action may be known by a craving, hankering, gone sensation before eating; by heart-burn, pain in the stomach, eructations, a dull, heavy, or painful sensation in the head, and a desire to be always nibbling at something: lives to eat, instead of eating to live, and should at once be eradicated by omitting one meal daily, and, in its stead, drinking abundantly of cold water.

LARGE. — Have a hearty relish for food; set a high value upon table enjoyments, and solid, hearty food; with Acquisitiveness large, lay up abundance of food for future use — perhaps keep so much on hand that some spoils; with Ideality large, must eat from a clean plate, and have food nicely cooked; with large Language and intellect, enjoy table-talk exceedingly, and participate in it; with large social faculties, must eat with others; can cook well, if practiced in culinary arts; and with larger Approbativeness and Ideality than Causality, are apt to be ceremonious and over-polite at table, etc. Such should restrain this faculty by eating less, more slowly, and seldom: p. 86.

FULL. — With a healthy stomach, eat freely what is offered, asking no questions; enjoy it, but not extravagantly; rarely overeat, except when the stomach is disordered, and then experience that hankering above described, which a right diet alone can cure. For combinations, see Large: p. 87.

AVERAGE. — Enjoy food well, and eat with a fair relish; yet rarely overeat, except when rendered craving by dyspeptic complaints: p. 86.

MODERATE. — Rather lack appetite; eat with little relish, and hence require to pamper and cultivate appetite by dainties, and enjoying rich flavors; can relish food only when other circumstances are favorable; feel little hunger, and eat to live, instead of living to eat; with Eventuality small, cannot remember from one meal to another what was eaten at the last: p. 87.

SMALL. — Eat "with long teeth," and little relish; hardly know or care what or when they eat; and should pay more attention to duly feeding the body: p. 88.

VERY SMALL. — Are almost wholly destitute of appetite.

This faculty is more liable to perversion than any other, and excessive and fast eating occasions more sickness, and depraves the animal
passions more than all other causes combined. Properly to feed the
body is of the utmost importance. Whenever this faculty becomes
diseased, the first object should be to restore its natural function by
right eating. Medicines rarely do it.

To CULTIVATE. — Consider before you provide or order your
meals what would relish best, and as far as possible provide what
seems to you will taste good; pamper appetite; eat leisurely, and as
if determined to extract from your food all the rich flavors it may contain, and in eating be governed more by flavor than quantity; endeavor to get up an appetite, even when you feel none, by eating
some dainty, as if to see if it were not good; do by food and drinks
as wine connoisseurs do in testing viands; taste things with a view
of ascertaining their relative flavors; in short, exercise and indulge
appetite; also, do as directed in order to cultivate digestion : 245.

To RESTRAIN. — Eat but seldom — for by keeping away from table
this faculty remains comparatively quiet; and when you eat, eat
slowly, leisurely, quietly, pleasurably, as if determined to enjoy eating, for this satisfies appetite with much less food than to eat vora-
ciously; mingle pleasant conversation with meals; direct attention
more to how good your food than how much you eat; always leave the
table with a good appetite, and stop the moment you have to resort
to condiments or desserts to keep up appetite; eat like the epicure,
but not like the gourmand — as if you would enjoy a little rather than
devour so much; eat sparingly, for the more you eat the more you re-
inflame the stomach, and thereby reincrease that hankering you need
to restrain : 246.

F. BIBATIVENESS OR AQUATIVENESS.
(Located in front of Alimentiveness.)

THE DRINKER, AND THE SWIMMER. — Love of liquids;
fondness for water, washing, bathing, swimming, sailing,
stimulants, etc. Adapted to the existence and utility of water.
Perversion — drinking in excessive quantities; drunkenness ;
and unquenchable thirst.

VERY LARGE. — Are excessively fond of water, whether applied
internally or externally, and a natural swimmer; and with Individuality and Locality, a natural seaman; with large Adhesiveness and

Approbativeness, ano small Self-esteem and Acquisitiveness, should avoid the social glass, for fear of being overcome by it.

LARGE. — Love to drink freely and frequently; experience much thirst; enjoy washing, swimming, bathing, etc., exceedingly, and are benefited by them; with Ideality large, love water prospects.

FULL. — Enjoy water well, but not extravagantly; drink freely when the stomach is in a fevered state, and are benefited by its judicious external application.

AVERAGE. — Like to drink at times, after working freely or perspiring copiously, yet ordinarily care little about it.

MODERATE. — Partake of little water, except occasionally, and are not particularly benefited by its external application, further than is necessary for cleanliness; dislike shower or plunge-baths, and rather dread than enjoy sailing, swimming, etc., especially when Cautiousness is large.

SMALL. — Care little for liquids in any of their forms, or for any soups, and, with large Cautiousness, dread to be on or near the water: with Alimentiveness large, prefer solid, hard food to puddings or broth, etc.

VERY SMALL. — Have an unqualified aversion to water and all fluids.

11. ACQUISITIVENESS.

LARGE.

SMALL.

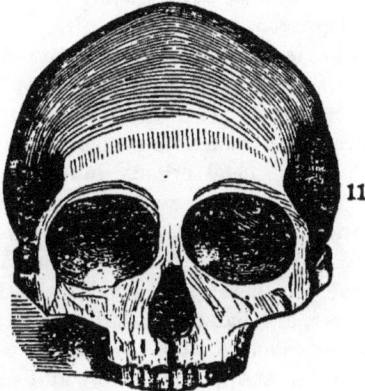

No. 152. — WM. TELLER, THIEF AND MURDERER.

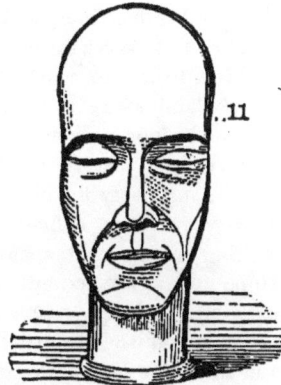

No. 153. — MR. GOSSE—GAVE AWAY TWO FORTUNES.

THE MERCHANT. — Economy; frugality; the acquiring saving, and laying-up instinct; laying up a surplus. and allowing nothing to be wasted; desire to possess and own; the mine-and-thine feeling; claiming of one's own things; love of trading and amassing property. Adapted to man's need of laying up the necessaries and comforts of life against future needs. Perversion — a miserly, grasping, close-fisted penuriousness.

VERY LARGE. — Hasten to be rich; are too eager after wealth; too industrious; too close in making bargains; too small in dealing; with moderate Hope and large Cautiousness, are penny wise but pound foolish; hold the sixpence too close to the eye to see the dollar farther off, and give entire energies to amassing property; with smaller Secretiveness and large Conscientiousness, are close, yet honest; will have dues, yet want no more, and never employ deception; but, with large Secretiveness and but average Conscientiousness, make money anyhow; palm off inferior articles for good ones, or at least over-praise those on sale, but run down in buying; and with large Parental Love and perceptives added, can make a finished horse-jockey; with small Self-esteem, are small and mean in deal, and stick for the half cent; with very large Hope and only full Cautiousness, embark too deeply in business, and are liable to fail; with large Adhesiveness and Benevolence, will *do* for friends more than give, and had rather *circulate* the subscription-paper than sign it; with large Hope and Secretiveness, and only average Cautiousness, buy more than can be paid for, pay more in promises than money, should adopt a cash business, and check the manifestations of this faculty by being less penurious and industrious, and more liberal: p. 92.

LARGE. — Save for future use what is not wanted for present; allow nothing to go to waste; turn everything to a good account; buy closely and make the most of everything; are industrious, economical, and vigorously employ all means to accumulate property; desire to own and possess much; with large social organs, industriously acquire property for domestic purposes, yet are saving in the family; with very large Adhesiveness and Benevolence, are industrious in making money, yet spend it too freely upon friends; with large Hope added, are too apt to indorse for them; with small Secretiveness, and activity greater than power, are liable to overdo, and take on too much work in order to save, so much as often to incur sickness, and thus lose more than gain; with large Approbativeness and small Secretiveness, boast of wealth, but with large Secretiveness, keep pecuniary affairs secret; with large Constructiveness, incline to make money by engaging in some mechanical branch of business; with large Cautiousness, are provident; with large Ideality, keep things very nice, and are tormented by whatever mars beauty; with large intellectual organs, love to accumulate books, and whatever facilitates intellectual progress; with large Veneration and Self-esteem, set great store by antique and rare coins, and specimens, etc.: p. 89.

FULL. — Take good care of possessions, and use vigorous exertions to enhance them; value property for itself and its uses; are industrious, yet not grasping; and saving, without being close; with large Benevolence, are too ready to help friends; and with large Hope added, too liable to indorse; and with an active temperament, too industrious to come to want; yet too generous ever to be rich. p. 93.

AVERAGE. — Love property; yet the other faculties spend quite as fast as this faculty accumulates; with Cautiousness large or very large, love property in order to be safe against future want; with large Approbativeness, desire it to keep up appearances; with large Conscientiousness, to pay debts; with large intellectual organs, will pay freely for intellectual attainments; yet the kind of property and objects sought in its acquisition depends upon other and larger faculties: p. 89.

MODERATE. — Value and make property more for its uses than itself; seek it as a means rather than an end; with Cautiousness large, may evince economy from fear of coming to want, or with other large organs, to secure other ends, yet care little for property on its own account; are rather wasteful; do not excel in bargaining, or like it; have no great natural pecuniary tact, or money-making capability, and are in danger of living quite up to income; with Ideality large, must have nice things, no matter how costly, yet do not take first-rate care of them; disregard small expenses; purchase to consume as soon as to keep; prefer to enjoy earnings now to laying them up; with large domestic organ, spend freely for family; with strong Approbativeness and moderate Cautiousness, are extravagant, and contract debts to make a display; with Hope large, run deeply in debt, and spend money before it is earned: p. 94.

SMALL. — Hold money loosely; spend it often without getting its value; care little how money goes; with Hope very large, enjoy money to-day without saving for to-morrow; and with large Approbativeness and Ideality added, and only average Causality, are prodigal, and spend money to poor advantage; contract debts without providing for their payment, etc.: p. 95.

VERY SMALL. — Neither heed nor know the value of money; **are** wasteful; spend all they can get; lack industry, and will be **always** in want: p. 95.

The back part of this organ, called Acquisition, accumulates property; the fore part, called Accumulation, saves; the former large and latter small, encompass sea and land to make a dollar, and then throw it away, which is an American characteristic; and get many things, but allow them to go to waste. Properly to spend money implies a high order of wisdom. Every dollar should be made an instrument of the highest happiness.

To CULTIVATE. — Try to estimate the value of money intellectually, and save up as a philosophy; economize time and means; cultivate industry; engage in some mercenary business; determine to get rich, and use the means for so doing, and be what you consider even small in expenditures; lay by a given sum at stated times, without thinking to use it except in extreme want; and when enough is laid by, make a first payment on real estate, or launch into business, thus compelling yourself both to save the driblets, and earn what you can in order to save yourself, and do by intellect what you are not disposed to do by intuition: p. 249.

To RESTRAIN. — Think less of dollars; study means for enjoying your property; often quit business for recreation; attend more relatively to other life ends, less to mere money-getting; that is, **cultivate** the other faculties, and be more generous: p. 250.

12. SECRETIVENESS.

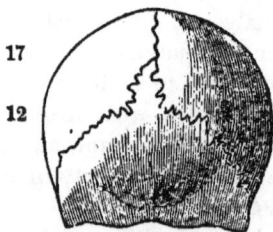   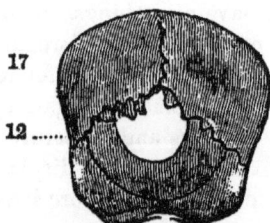

No. 154. — LARGE.   No. 155. — SMALL.

THE CONCEALER. — Tact; self government; ability to re **strain** feelings; policy; management; reserve; evasion; **dis** cretion; cunning; double-dealing; adapted to man's requisition for self-control. Perverted, it causes duplicity, double-dealing, lying, deception, and all kinds of false pretensions.

It is located in the centre of the animal group, doubtless because we require to restrain our passions most.

VERY LARGE. — Are non-committal and cunning in the extreme, and with only average Conscientiousness, deceptive, tricky, foxy, double-dealing, and unworthy to be trusted; with large Acquisitiveness added, will both cheat and lie; with large Cautiousness, are unfathomable even by acknowledged friends; with very large moral organs, and only average or full propensities, are not dangerous, and have a good moral basis, yet instinctively employ many stratagems calculated to cover up the real motives; and should cultivate openness and sincerity: p. 98.

LARGE. — Throw a veil over countenance, expression, and conduct; appear to aim at one thing, while accomplishing another; love to surprise others; are enigmatical, mysterious, guarded, politic, shrewd, managing, employ humbug, and are hard to be found out; with Cautiousness large, take extra pains to escape detection; with Conscientiousness also large, will not tell a lie, yet will not always tell the truth; evade the direct question, and are equivocal, and though honest in purpose, yet resort to many little cunning devices; with large intellectual organs and Cautiousness, express ideas so guardedly as to lack distinctness and directness, and hence to be often misunderstood; with large Approbativeness, take many ways to secure notoriety, and hoist some false colors; with large Acquisitiveness, employ too much cunning in pecuniary transactions, and unless checked by still larger Conscientiousness, are not always strictly truthful or honest; with large social organs, form few friendships, and those only after years of acquaintance, nor evince half the attachment felt; are distant in society, and communicate even with friends only by piecemeal; divulge very few plans or business matters to acquaintances, or even to friends; lack communicativeness, and have little or no fresh-hearted expression of feeling, but leave an impression of uncertainty as to character and intention: p. 96.

FULL. — Evince much self-government; yet, if the temperament is active, when the feelings do break forth, manifest them with unusual intensity; with large Acquisitiveness and Cautiousness, communicate but little respecting pecuniary affairs; with large Approbativeness,

7

take the popular side of subjects, and sail only with the current of public opinion; with Conscientiousness large, are upright in motive, and tell the truth, but not always the whole truth; and though never hoist false colors, yet do not always show true ones: p. 99.

AVERAGE. — Maintain a fair share of self-government, except when under excitement, and then let the whole mind out fully; with large Combativeness and an active temperament, though generally able to control resentment, yet, when once provoked, show the full extent of their anger; with large Cautiousness, see that there is no danger before allowing the feelings to burst forth; but with an excitable temperament, and especially a deranged stomach, show a general want of policy and self-government, because the feelings are too strong to be kept in cheek; but if this faculty is manifested in connection with stronger faculties, it evinces considerable power, yet is wanting when placed in opposition to them: p. 96.

MODERATE. — Express feelings with considerable fullness; pursue an open, direct course; are sincere and true; employ but little policy, and generally give full vent to thoughts and feelings; with Cautiousness large, evince prudence in deeds, but imprudence in words; express opinions unguardedly, yet are safe and circumspect in conduct; with large Acquisitiveness and Conscientiousness, are honest, and think others equally so; are too easily victimized by the confidence man; prefer the one-price system in dealing, and cannot bear to banter; with large Adhesiveness, are sincere and open-hearted in friendship, and communicate with perfect freedom; with large Conscientiousness and Combativeness added, are truthful, and speak the whole mind too bluntly; with fine feelings, and a good moral organization, manifest the higher, finer feelings without restraint or reserve, so as to be the more attractive; are full of goodness, and show all that goodness without any intervening veil; manifest in looks and actions what is passing within; express all mental operations with fullness, freedom, and force; choose direct and unequivocal modes of expression; disclose faults as freely as virtues, and leave none at a loss as to the real character; but with the harsher elements predominant, appear more hating and hateful than they really are, because all is blown right out: p. 100.

SMALL. — Are perfectly transparent; seem to be just what, and all they really are; disdain concealment in all forms; are no hypocrites, but positive and unequivocal in all said and done; carry the soul in the hands and face, and make way directly to the feelings of others, because expressing them so unequivocally; are too spontaneous; with large Cautiousness, are guarded in action, but unguarded in expression; free the mind regardless of consequences, yet show much prudence in other respects; with Conscientiousness large, love the truth wherever it exists, and open the mind freely to evidence and conviction; are open and aboveboard in everything, and allow all the mental operations to come right out, unveiled and unrestrained, so that their full force is seen and felt: p. 101.

VERY SMALL. — Conceal nothing, but disclose everything: p.

TO CULTIVATE. — Supply by intellect that guardedness and policy lacked by instinct; try to "lie low, and keep dark," and suppress your natural outgushings of feeling and intellect; cultivate self-control by subjecting all you say and do to judgment, instead of allowing momentary impulses to rule conduct; do not tell all you know or intend to do, and occasionally pursue a roundabout course; be guarded, politic, and wary in everything; do not make acquaintances or confide in people as much as is natural, but treat everybody as if they needed watching: 252.

TO RESTRAIN. — Cultivate a direct, straightforward, aboveboard, and open way, and pursue a course just the opposite from the one suggested for its cultivation: 253.

13. CAUTIOUSNESS.

THE SENTINEL. — Carefulness; prudence; solicitude; anxiety; watchfulness; apprehension; security; protection; provision against want and danger; foreseeing and avoiding prospective evils; the watchman; discretion; care; vigilance.

Adapted to ward off surrounding dangers, and make those provisions necessary for future happiness. Perversion — irresolution; timidity; procrastination; indecision; fright; panic.

VERY LARGE. — Are over-anxious; always on the lookout; worried about trifles; afraid of shadows; forever getting ready, because so many provisions to make; are careful in business; often revise decisions, because afraid to trust the issue; live in perpetual fear of evils and accidents; take extra pains with everything; lack promptness and decision, and run no large risks; put off till to-morrow what ought to be done to-day; with excitability large, live in a constant panic; procrastinate; are easily frightened; see mountains of evil where there are only mole-hills; are often unnerved by fright, and overcome by false alarms; with only average or full Combativeness, Self-esteem, and Hope, and large Approbativeness, accomplish literally nothing,

LARGE.   SMALL.

No. 156. — DEACON TERRY.   No. 157. — CHARLES XII. OF SWEDEN.

but should always act under others; with large Acquisitiveness, pre-fer bonds and mortgages to traffic, small but sure gains to large but more risky ones, and safe investments to active business: p. 105.

LARGE. — Are always on the lookout; take ample time to get ready; provide against prospective dangers; make everything safe; guard against losses and evils; incur no risks; sure bind that they may sure find; with large Combativeness, Hope, and an active tem-perament, drive, Jehu-like, whatever is undertaken, yet drive cau-tiously; lay on the lash, yet hold a tight rein, so as not to upset plans; with large Approbativeness, are doubly cautious as to charac-ter; with large Approbativeness and small Acquisitiveness, are extra careful of character, but not of money; with large Acquisitiveness and small Approbativeness, take special care of all money matters, but not of reputation; with large Adhesiveness and Benevolence, experience the greatest solicitude for the welfare of friends; with large Consci-entiousness, are careful to do nothing wrong; with large Causality, ay safe plans, and are judicious; with large Combativeness and Hope, combine judgment with energy and enterprise, and often seem reck-less, yet are prudent; with large intellectual organs and Firmness, are cautious in coming to conclusions, and canvass well all sides of all questions, yet, once settled, are unmoved; with small Self-esteem, rely too much on the judgment of others, and too little on self; with large Parental Love and disordered nerves, experience unnecessary solicitude for children, and take extra care of them, often killing them with kindness, etc.: p. 104.

FULL.— Show a good share of prudence and carefulness, except when the other faculties are powerfully excited; with large Combativeness and very large Hope, have too little prudence for energy; are tolerably safe, except when under considerable excitement; with large Acquisitiveness, are very careful whenever money or property are concerned; yet, with only average Causality, evince but little general prudence, and lay plans for the present rather than future, etc.: p. 105.

AVERAGE. — Have a good share of prudence, whenever this faculty works in connection with the larger organs, yet evince but little in the direction of the smaller; with large Combativeness and Hope, and an excitable temperament, are practically imprudent, yet somewhat less so than appearances indicate; with large Causality and only average Hope and Combativeness, and a temperament more strong than excitable, evince good general judgment, and meet with but few accidents; but with an excitable temperament, large Combativeness and Hope, and only average or full Causality, are always in hot water, fail to mature plans, begin before ready, and are luckless and unfortunate in everything, etc.: p. 103.

MODERATE. — With excitability great, act upon the spur of the moment, without due deliberation; meet with many accidents caused by imprudence; with large Combativeness, are often at variance with neighbors, and make many enemies; with large Approbativeness, seek praise, yet often incur criticism; with average Causality and large Hope, are always doing imprudent things, and require a guardian; with small Acquisitiveness, keep money loosely, and are easily over-persuaded to buy more than can be paid for; with large Parental Love, play with children, yet often hurt them; with large Language and small Secretiveness, say many very imprudent things, etc.: p. 106.

SMALL.— Are rash, reckless, luckless; and with large Hope, always in trouble; with large Combativeness, plunge headlong into difficulties in full sight, and should assiduously cultivate this faculty: p. 106.

VERY SMALL. — Have so little of this faculty, that its influence on conduct is rarely ever perceived: p. 107.

To Cultivate. — Count the advantages against, but not for look out for breakers; think how much indiscretion and carelessness have injured you, and be careful and watchful in everything. Imprudence is your fault. Be judicious; and remember that danger is always much greater than you anticipate; so keep aloof from every appearance of it: 255.

To Restrain. — Offset its workings by intellect; remember that you perpetually magnify dangers; let intellect tell Cautiousness to keep quiet; offset it by cultivating a bold, combative, daring spirit; encourage a don't-care feeling, and a let-things-take-their-course — why-worry-about-them? do not indulge so much anxiety when children or friends do not return as expected; never allow a frightened, panic-stricken state of mind, but face apprehended evils, instead of quailing before them; and remember that you magnify every appearance of evil: 256.

14. APPROBATIVENESS.

The Aristocrat. — Ambition; regard for character, appearances, etc.; love of praise, popularity, fashion, and fame; desire to excel and be esteemed; affability; politeness; love of display and show; sense of honor; desire for a good name, for notoriety, eminence, distinction, and to be thought well of; pride of character; sensitiveness to the speeches of people.

158.—The Proud Youth.

Adapted to the reputable and disgraceful. Perversion — vanity; affectation; ceremoniousness; aristocracy; pomposity; eagerness for popularity, gaudy display, etc.

Very Large. — Set everything by the good opinion of others; are ostentatious, if not vain and ambitious; love praise, and are mortified by censure inordinately; with moderate Self-esteem and Firmness, cannot breast public opinion, but are over-fond of popularity; with only average Conscientiousness, seek popularity without regard to merit; but with large Conscientiousness, seek praise mainly for virtuous doings; with large Ideality, and only average Causality, seek

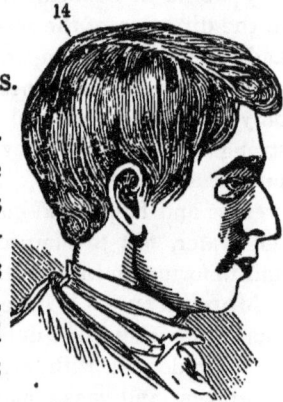

praise for fashionable dress and outside appearances rather than internal merit; are both vain and fashionable as well as aristocratic; starve the kitchen to ornament the parlor; with large Acquisitiveness, boast of riches; with large Adhesiveness, of friends; with large Language, are extra forward in conversation, and engross much of the time, etc. This is the main organ of aristocracy, exclusiveness, fashionableness, so-called pride, and nonsensical outside show: p. 110.

LARGE. — Love commendation, and are cut by censure; are keenly alive to the smiles and frowns of public opinion; mind what people say; strive to show off to advantage, and are affable, courteous, and desirous of pleasing; love to be in company; stand on etiquette and ceremony; aspire to do and become something great; set much by appearances, and are mortified by reproach; with large Cautiousness and moderate Self-esteem, are bashful, take the popular side, and fear to face the ridicule of others; yet, with Conscientiousness and Combativeness large, stick to the right, though unpopular, knowing that it will ultimately confer honor; with large Benevolence, seek praise for works of philanthropy and mercy; with large intellectual organs, love literary and intellectual distinctions; with large Adhesiveness, desire the good opinion of friends, yet care little for that of others; with large Self-esteem, Combativeness, and excitability, are very touchy when criticised, seek public life, want all the praise, and hate rivals; with large perceptives, take a forward part in literary and debating societies; with large Combativeness, Hope, and activity, will not be outdone, but rather work till completely exhausted, and are liable to overdo, in order to eclipse rivals: p. 108.

FULL. — Value the estimation of others, yet will not go far after it; seek praise in the direction of the larger organs, yet care little for it in that of the smaller; are not aristocratic, yet like to make a fair show in the world; with large Adhesiveness, seek the praise and avoid the censure of friends; with large Conscientiousness, set much by *moral* character, and wish to be praised for correct *motives*; yet, with moderate Acquisitiveness, care little for the name of being rich; with large Benevolence and intellectual organs, desire to be esteemed for evincing talents in doing good, etc.: p. 110.

AVERAGE. — Show only a respectable share of this faculty, except when it is powerfully wrought upon by praise or reproach: are mortified by censure, yet not extremely so, and call in the other faculties to justify; are not particularly ambitious, yet by no means deficient, and not insensible to compliments, yet cannot well be inflated by praise: p. 107.

MODERATE. — Feel some, but no great, regard for popularity and evince this faculty only in connection with the larger organs with large Self-esteem and Firmness, are inflexible and austere; and with large Combativeness and small Agreeableness, lack civility and complaisance to others; disdain to flatter, and cannot be flattered, and should cultivate a pleasing, winning mode of address : p. 112.

SMALL. — Care little for the opinions of others, even of friends; are comparatively insensible to praise ; disregard style and fashion ; despise etiquette and formal usages; never ask what will persons think, and put on no outside appearances for their own sake ; with large Self-esteem, Firmness, and Combativeness, are destitute of politeness, devoid of ceremony, and not at all flexible or pleasing in manners; with large Combativeness and Conscientiousness, go for the right regardless of popularity, and are always making enemies; say and do things in so graceless a manner as often to displease; with large Acquisitiveness and Self-esteem, though wealthy, make no boast of it, and are as commonplace in conduct as if poor, etc. : p. 112.

VERY SMALL. — Care almost nothing for reputation, praise, or censure.

TO CULTIVATE. — Remember that you often stand in your own light by caring too little for the speeches of people, for appearance and character ; and cherish a higher regard for public opinion, for your character and standing among men, for a good name, and do nothing in the least to tarnish your reputation, but cultivate a winning, politic, pleasant manner toward all, as if you would ingratiate yourself into their good-will : 258.

TO RESTRAIN. — Remember that you are infinitely too sensitive to reproof; that your feelings are often hurt when there is no occasion ; that you often feel neglected or reproved without cause; that evil speaking breaks no bones, and will ultimately thwart itself; should lay aside that affected, artificial style of manners and speaking ; be more natural ; walk, act, feel as if alone, not forever looked at; be less particular about dress, style, appearance, etc., and less mindful of praise and blame ; subject Approbativeness to conscience ; that is, do what is right, and let people say what they like ; be more independent, and less ambitious and sensitive to praise and flattery 259.

15. SELF-ESTEEM.

THE IMPERATOR. — Self-respect, self-reliance, self-apprecia-
tion, self-satisfaction, and complacency; independence; dig-
ty; nobleness; love of liberty and power; the aspiring,
self-elevating, ruling instinct.

Adapted to the superiority, greatness, and exalted dignity of
human nature.   Perversion — egotism; hauteur; forwardness;
vranny; superciliousness; imperiousness.

VERY LARGE. — Have the highest respect for *self;* place special
tress on the personal pronouns; carry a high head, and walk so
straight as to lean backward; have a restless, boundless ambition to
e and do some great thing; with only full intellect, have more egotism than talents, and are proud, pompous, supercilious, and imperious, and with Hope large, must operate on a great scale or none, and
launch out too deeply; with Approbativeness large, are most aristocratic; and with only fair intellect, are a swell-head and great brag,
and put self above everybody else; with only average Approbativeness and Agreeableness, take no pains to smooth off the rougher
points of character, but are every way repulsive; with average Parental Love, are very domineering in the family, and insist upon being
waited upon, obeyed, etc.; and should carry the head a little lower,
and cultivate humility: p. 116.

LARGE. — Put a high estimate upon own sayings, doings, and
capabilities; fall back upon own unaided resources; will not take advice, but insist upon being own master; are high-minded; will never
stoop, or demean self: aim high; are not satisfied with moderate success, or a petty business, and comport and speak with dignity, perhaps majesty; are perfectly self-satisfied; with large Parental Love,
pride self in children, yet with Combativeness large, require implicit
obedience, and are liable to be stern; with large Adhesiveness, seek
society, yet must lead; with large Acquisitiveness added, seek part
nership, but must be the head of the firm; with large Firmness and
Combativeness, cannot be driven, but insist upon doing *own* will and
pleasure, and are sometimes contrary and headstrong; with large
Hope, think that anything you do must succeed, because done so
well; with large moral organs, impart a tone, dignity, aspiration, and
elevation of character which command universal respect; and with

large intellectual faculties added, enjoy and are very well calculated for public life; are a natural leader, but seek moral distinction, and to lead the public *mind;* with large Combativeness, Destructiveness, Firmness, and Approbativeness, love to be captain or general, and speak with that sternness and authority which enforce obedience; with large Acquisitiveness, aspire to be rich — the richest man in town — partly on account of the power wealth confers; with large Language, Individuality, Firmness, and Combativeness, seek to be a political leader; with large Constructiveness, Perceptives, Causality, and Combativeness, are well calculated to have the direction of men, and oversee large mechanical establishments; with only average brain and intellect, and large selfish faculties, are proud, haughty, domineering, egotistical, overbearing, greedy of power and dominion, etc.: p. 114.

FULL. — Evince a good degree of dignity and self-respect, yet are not proud or haughty; with large Combativeness, Firmness, and Hope, rely fully upon own energies in cases of emergency, yet are willing to hear advice, though seldom take it; conduct becomingly, and secure respect; and with large Combativeness and Firmness, and full Destructiveness and Hope, evince much power of this faculty, but little when these faculties are moderate: p. 116.

AVERAGE. — Show this faculty mainly in combination with those that are larger; with large Approbativeness and Firmness, and a large brain and moral organs, rarely trifle or evince meanness, yet are rarely conceited, and think neither too little nor too much of self, but place a just estimate upon their own capabilities; with large Adhesiveness, both receive and impart character to friends, yet receive most; with large Conscientiousness, pride self more on moral worth than physical qualities, wealth, titles, etc.; and with large intellectual and moral organs, mainly for intellectual and moral excellence: p. 112.

MODERATE. — Rather underrate personal capabilities and worth; feel somewhat inferior, unworthy, and humble; lack dignity, and are apt to say and do trifling things, and let self down; with large intellectual and moral organs, lead off well when once placed in a responsible position, yet at first distrust own capabilities; with large Conscientiousness, Combativeness, and activity, often appear self-sufficient and positive, because certain of being right, yet it is founded more on reason than egotism; with large Approbativeness, love to show off, yet are not satisfied with self; go abroad after praise, rather than feel internally conscious of personal merits; are apt to boast, because more desirous of the estimation of others than conscious of persona worth; with large moral and intellectual powers, have exalted thoughts and aspirations, and communicate well, yet often detract from them by commonplace phrases and undignified expressions; will be too familiar to be respected in proportion to merit, and should vigorously cultivate this faculty by banishing mean, and cultivating high thoughts of self: p. 116.

SMALL. — Feel diminutive; lack elevation and dignity of tone and manner; place too low estimate on self; and, with Approbativeness large, are too anxious to appear well in the eyes of others; with large Combativeness and Destructiveness, show some self-reliance when provoked or placed in responsible positions, yet lack that dignity which commands respect, and leads off in society; lack self-confidence and weight of character; shrink from responsible and great undertakings, from a feeling of unworthiness; underrate self, and are therefore undervalued by others, and feel insignificant, as if in the way, or trespassing upon others, and hence often apologize, and should cultivate this faculty.

VERY SMALL. — Feel little, and manifest none of this faculty.

To CULTIVATE. — Say of yourself what Black Hawk said to Jackson — "I am a man!" one endowed with the ennobling elements of humanity. Realize how exalted those human endowments conferred on you are, and put a higher estimate on yourself, physically, intellectually, morally. Recount your good traits, and cultivate self-valuation in view of them. Pride yourself on what you are, but never indulge self-abasement because not dressed, or not as rich or stylish as others. Be less humble toward men, but hold up your head among them, as if good enough for any. Assume the attitude and natural language of self-esteem. Study its phrenological definition, and cultivate the self-esteem feeling: 261.

To RESTRAIN. — Bear in mind that you esteem yourself much better than you really are; that you overrate all your powers, and are too forward and self-confident; that more modesty would improve yo. ; that you incline too much to be arbitrary and domineering; that you are more faulty than you suppose, and need humility: 263.

16. FIRMNESS.

THE PILLAR. — Stability; decision; perseverance; perti·
nacity; fixedness of purpose; aversion to change; indomita·
bility; will.

Adapted to man's requisition for holding out to the end.
Perversion — obstinacy; willfulness; mulishness; stubborn·
ness; unwillingness to change even when reason requires.

VERY LARGE. — Are well-nigh obstinate, stubborn, and with large
Combativeness and Self-esteem, as unchangeable as the laws of the
Medes and Persians, and can neither be persuaded nor driven; with
large activity, power, brain, and intellectual organs, are well calculated to carry forward some great work which requires the utmost determination and energy; with large Causality, can possibly be turned
by potent reasons, yet by nothing else.

LARGE. — Are set and willful; stick to and carry out what is commenced; hold on long and hard; continue to the end, and may be
fully relied upon; with full Self-esteem and large Combativeness, cannot be driven, but the more determined the more driven; with large
Combativeness and Destructiveness, add perseverance to stability,
and not only hold on, but drive forward determinedly through difficulties; with large Hope, undertake much, and carry all out; with
large Cautiousness and Causality, are careful and judicious in laying
plans and forming opinions, yet rare-
ly change; may seem to waver until
the mind is fully made up, but are
afterward the more unchanging; with
Hope very large, and Cautiousness
and Causality only average, decide
quickly, even rashly, and refuse to
change; with Adhesiveness and Benevolence large, are easily persuaded,
especially by friends, yet cannot be
driven; and with large Cautiousness,
Combativeness, Causality, perceptives, activity, and power, will generally succeed, because wise in planning and persevering in execu-

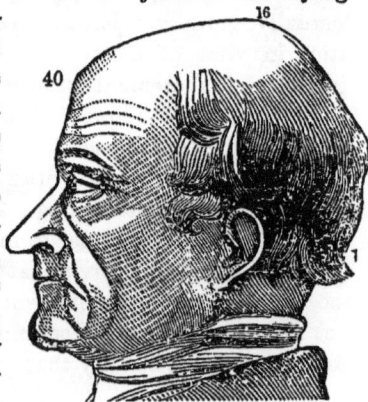

159.—DR. CALDWELL.   VERY LARGE.

tion; with Combativeness and Self-esteem large, and Causality only average, will not see the force of opposing arguments, but tenaciously adhere to affirmed opinions and purposes; with large Conscientiousness and Combativeness, are doubly decided wherever right and justice are concerned, and in such cases will never give one inch, but will stand out in argument, effort, or as juryman, till the last : p. 119.

FULL. — Like Firmness large, show a great degree of decision when this faculty works with large organs, but not otherwise; with Combativeness and Conscientiousness large, show great fixedness where right and truth are concerned, yet with Acquisitiveness moderate, lack perseverance in money matters; with moderate Combativeness and Self-esteem, are easily turned ; and with large Adhesiveness and Benevolence, too easily persuaded, even against better judgment; with Cautiousness and Approbativeness large, or very large, often evince fickleness, irresolution, and procrastination; and with an uneven head, and an excitable temperament, often appear deficient in this faculty : p. 131.

AVERAGE. — When supported by large Combativeness, or Conscientiousness, or Causality, or Acquisitiveness, etc., show a good degree of this faculty; but when opposed by large Cautiousness, Approbativeness, or Adhesiveness, evince its deficiency, and have not enough for great undertakings ; p. 119.

MODERATE. — Rather lack perseverance, even when the stronger faculties support it; but when they do not, evince fickleness, irresolution, indecision, and lack perseverance; with Adhesiveness large, are too easily persuaded and influenced by friends; with large Cautiousness and Approbativeness, and moderate or small Self-esteem, are flexible and fickle, and go with the current: p. 132.

SMALL. — With activity great, and the head uneven, are fitful, impulsive, and, like the weather-vane, shift with every changing breeze, and are ruled by the other faculties, and as unstable as water : p. 122.

VERY SMALL. — Are changed by the slightest motives, and a perfect creature of circumstances, and accomplish nothing requiring perseverance : p. 122.

TO CULTIVATE. — Have more a mind of your own : make up your mind wisely, and then stand to your purpose; be sure you are right, then hold on ; surmount difficulties, instead of turning aside to avoid them ; resist the persuasions of others; begin nothing not worthy of finishing, and finish all you begin : 265.

TO RESTRAIN. — Remember that you are too obstinate and persistent, often to your own loss; at least listen to the advice of others, and duly consider it, and govern Firmness by Intellect and Conscience, not allow it to govern them: 266.

MORAL SENTIMENTS.

These render man a moral, accountable, and religious being, humanize, adorn, and elevate his nature; connect him with the moral nature of things; create his higher and nobler faculties; beget aspirations after goodness, virtue, purity, and moral principle, and ally him to angels and to God.

VERY LARGE. — Have a most exalted sense and feeling of the moral and religious, a high order of practical goodness, and the strongest aspirations for a higher and holier state, both in this life and that which is to come.

LARGE. — Experience a high regard for things sacred and religious; have an elevated moral and aspiring cast of feeling and conduct, along with right intentions, and a strong desire to become good, holy, and moral in feeling and conduct; and with weak animal feelings, are too good for own good.

FULL. — Have good moral and religious tone, and general correctness of motive, so as to render feelings and conduct about right; but with strong propensities, and only average intellectual faculties, are

No. 160. — REV. DR. TYNG.   No. 161. — HAGARTY, MURDERER.

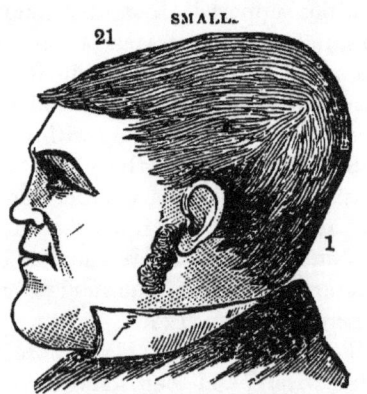

sometimes led into errors of belief and practice; mean right, yet sometimes do wrong, and should cultivate these faculties, and restrain the propensities.

AVERAGE. — Surrounded by good influences, will be tolerably moral and religious in feeling, yet not sufficiently so to withstand strong propensities; with disordered nerves, are quite liable to say and do wrong things, yet afterward repent, and require much moral cultivation.

MODERATE. — Have a rather weak moral tone; feel but little regard for things sacred and religious; are easily led into temptation; feel but little moral restraint; and, with large propensities, especially if circumstances favor their excitement, are exceedingly liable to say and do what is wrong.

SMALL. — Have weak moral feeling; lack moral character; and with large organs of the propensities, are liable to be depraved, and a bad member of society.

VERY SMALL. — Feel little, and show no moral tendencies.

To CULTIVATE. — Yield implicit obedience to the higher, better sentiments of your nature; cultivate a respect for religion; lead a moral, spotless life; cultivate all the human virtues; especially study and contemplate Nature, and yield yourself to those elevating influences kindled thereby; cultivate adoration and love of the Deity in His works; obey His natural laws; study natural religion, and make your life as pure, right, true, and good, as possible.

To RESTRAIN. — To avoid becoming morbid in the action of the moral sentiments, and to overrule it when it exists, subject Benevolence, justice, Veneration, devotion, and Spirituality, to the guidance of intellect; and be more selfish, or at least less self-sacrificing, and think more of material things.

17. CONSCIENTIOUSNESS.

THE JURIST. — Integrity; moral rectitude and principle, love of right and truth; regard for duty, moral purity, promises, and obligations; penitence; contrition; approval of right; condemnation of wrong; obedience to laws, rules, etc. Adapted to natural right and wrong, and to the natural laws, and the moral nature and constitution of things. Perverted, it makes one do wrong from conscientious scruples, and torments with undue self-condemnation.

VERY LARGE. — Place moral excellence at the head of all excellence; make duty everything; are governed by the highest order of moral principle; would on no account knowingly do wrong; are scrupulously exact in all matters of right; perfectly honest in motive;

17   16   17   17   16   17

No. 162. — VERY LARGE.   No. 163. — VERY SMALL.

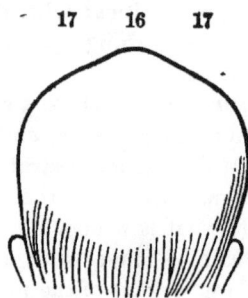

always condemning self and repenting, and very forgiving to those who evince penitence, but inexorable without; with Combativeness large, evince the utmost indignation at the wrong, and pursue the right with great energy; are censorious, make but little allowance for the faults and follies of mankind, show extraordinary moral courage and fortitude; and are liable to denounce evil-doers; with large Friendship, cannot tolerate the least thing wrong in friends, and are liable to reprove them; with large Parental Love, exact too much from children, and with large Combativeness, are too liable to blame them; with large Cautiousness, are often afraid to do, for fear of doing wrong; with large Veneration, reasoning faculties, and Language, are a natural theologian, and take the highest pleasure in reasoning and conversing upon all things having a moral and religious bearing; with Veneration average, and Benevolence large or very large, cannot well help being a thorough-going reformer, etc.: p. 129.

LARGE. — Love the right as right, and hate the wrong because wrong; are honest, faithful, upright in motive; mean well; consult duty before expediency; feel guilty when conscious of having done wrong; ask forgiveness for the past, and try to do better in future; with strong propensities, will sometimes do wrong, but be exceedingly sorry therefor; and, with a wrong education added, are liable to do wrong, thinking it right, because these propensities warp conscience, yet mean well; with large Cautiousness, are solicitous to know what is right, and careful to do it; with weaker Cautiousness, sometimes do wrong carelessly or indifferently, yet afterward repent it; with large Cautiousness and Destructiveness, are severe on wrong-doers, and unrelenting until they evince penitence, but then cordially forgive; with large Approbativeness, keep the moral character pure and spotless, value others on their morals more than wealth, birth, etc., and make the word the bond; with large Benevolence, Combativeness, and Destructiveness, feel great indignation and severity against oppressors, and those who cause others to suffer by wronging them: with large Ideality, have strong aspirations after moral purity and excellence; with large reasoning organs, take great pleasure, and show much talent in reasoning upon and investigating moral subjects, etc.: p. 126.

FULL. — Have good conscientious feelings, and correct general intentions, yet are not quite as correct in action as intentions; mean well, yet with large Combativeness, Destructiveness, Amativeness, etc., may sometimes yield to these faculties, especially if the system is somewhat inflamed; with large Acquisitiveness, make very close bargains, and will take such advantages as are common in business, yet do not intend to wrong others out of their just dues, still, have more regard for money than justice; with large intellectual organs, love to reason upon subjects where right and duty are involved, yet too often take the ground of expediency, and fail to allow right its due weight; and should never allow conscience to be in any way weakened, but cultivate it assiduously: p. 130.

AVERAGE. — When not tempted by stronger faculties, will do what is about right; generally justify self, and do not feel particularly indignant at the wrong, or commendatory of the right; with large Approbativeness and Self-esteem, may do the honorable thing, yet where honor and right clash, will follow honor; with only average Combativeness and Destructiveness, allow many wrong things to pass unrebuked, or even unresented, and show no great moral indignation or force; with moderate or small Secretiveness and Acquisitiveness, and large Approbativeness, Benevolence, and Ideality, will do as nearly right, and commit as few errors as those with Secretiveness, Acquisitiveness, and Conscientiousness all large, and may be trusted, especially on honor, yet will rarely feel guilty, and should never be blamed, because Approbativeness will be mortified before conscience is convicted; with large propensities, especially Secretiveness and Acquisitiveness, and only full Benevolence, are selfish; should be dealt with

8

cautiously, and thoroughly bound in writing, because liable to be slip-
pery, tricky, etc.; and should cultivate this faculty by never allowing
the propensities to overcome it, and by always considering things in
the moral aspect: p. 124.

MODERATE. — Have some regard for duty in feeling, but less in
practice; justify self; are neither very penitent nor forgiving; even
temporize with principle, and sometimes let interest rule duty : p. 131.

SMALL. — Have few conscientious scruples, and little penitence,
gratitude, or regard for moral principle, justice, duty, etc., and are
governed mainly by the larger faculties; with large propensities, and
only average Veneration and Spirituality, evince a marked deficiency
of moral principle; with moderate Secretiveness and Acquisitiveness, and only full Destructiveness and Combativeness, and large
Adhesiveness, Approbativeness, Benevolence, Ideality, and intellect,
and a fine temperament, may live a tolerably blameless life; yet, on
close scrutiny, will lack the moral in feeling, but may be safely trusted,
because true to promises, that is, conscience having less to contend
with, its deficiency is less observable. Such should most earnestly
cultivate this faculty : p. 132.

VERY SMALL. — Are almost wholly destitute of moral feeling, and
wholly controlled by the other faculties: p. 133.

To CULTIVATE. — Always ask yourself what is right and wrong,
and adhere closely to the former, and studiously avoid the latter;
make everything a matter of principle; do just as nearly right as you
know how in everything, and never allow conscience to be borne
down by any of the other faculties, but keep it supreme; maintain
the right everywhere and for everybody; cultivate a high sense of
duty and obligation, and try to reform every error; in short, "let justice be done, though the heavens fall:" 268.

To RESTRAIN. — Remember that you are too exact and exacting
in everything; that you often think you see faults where there are
none; that you carry duty and right to a needless extreme, and so far
as to make it wrong; that you are too condemnatory, and need to
cultivate a lenient, forbearing, forgiving spirit; that you trouble yourself unduly about the wrong-doing of others; that you often accuse
people of meaning worse than they really intend, and look at minor
faults as mountains of wrong; are too censorious; too apt to throw
away the gold on account of dross, to discard the greater good on account of lesser attendant evils; too liable to feel guilty and unworthy,
as if unfit to live, and too conscience-stricken. Extreme Consci-

entiousness, with 6 or 7 organic quality, and large Combativeness, along with disordered nerves or dyspepsia, makes one of the most unpleasant of characters — querulous, eternally grumbling about nothing, magnifying everybody's faults, thus making mischief among neighbors; perpetually accusing everybody, and chiding children for mere trifles ; too rigid in matters of reform, and violent in denouncing opponents. of whom rabid radicals, punctilious religionists, and old maids furnish examples : 270.

18. HOPE.

The Promiser. — Anticipation of future success and happiness ; buoyancy ; light-heartedness ; that which looks on the bright side, builds fairy castles, magnifies prospects, speculates, makes promises, etc.

Adapted to man's relations with the future.   Perverted, it becomes visionary.

Very Large. — Have unbounded expectations ; build a world of castles in the air; live in the future ; enjoy things in anticipation more than possession ; with small Continuity, have too many irons in the fire; with an active temperament added, take on more business than can be worked off properly ; are too much hurried to do things in season ; with large Acquisitiveness, are grasping, count chickens before they are hatched, and often two to the egg at that; are always rushing on after great piles of money away ahead, without noticing the smaller sums near by ; with only average Cautiousness, are always in hot water ; never stop to enjoy what is possessed, but grasp after more, and would accomplish much more if less were undertaken, and in taking one step forward, often slip two steps back: p. 133.

Large. — Expect much from the future ; contemplate with pleasure the bright features of life's picture; never despond; overrate prospective good, and underrate and overlook obstacles and evils; calculate on more than the nature of the case will warrant; expect, and hence attempt, a great deal, and are therefore always full of business ; are sanguine, and rise above present trouble by hoping for better things in future, and though disappointed, hope on still ; build some air-castles, and live in the future more than present ; with large Combativeness, Firmness, and Causality, are enterprising, never give up the ship, but struggle manfully through difficulties ; and with large

Approbativeness, and full Self-esteem added, feel adequate to difficulties, and grapple with them spiritedly; with large Self-esteem,
think everything I attempt must succeed, and with large Causality
added, consider their plans well-nigh perfect; with large Acquisitiveness, lay out money freely in view of future gain; with large Approbativeness and Self-esteem, hope for renown, honor, etc.; with large
Veneration and Spirituality, hope to attain exalted moral excellence,
and should check it by acting on only half it promises, and reasoning
against it: p. 137.

FULL. — Expect considerable, but undertake no more than can be
accomplished; are quite sanguine and enterprising, yet with Cautiousness large are always on the safe side; with large Acquisitiveness added, invest money freely, yet always safely, and belong to the
" bears;" make good bargains, if any, and count all the cost, yet are
not afraid of expenses where they will more than pay; with larger
animal organs than moral, will hope more for this world's goods than
for another's, and with larger moral than animal, for happiness in
another state of being than in this, etc.: p. 139.

AVERAGE. — Expect and attempt too little, rather than too much,
with large Cautiousness, dwell more on difficulties than encourage-
ments; are contented with the present rather than lay up for the future; with large Acquisitiveness added, invest money very safely, if
at all, and prefer to put it out securely on interest rather than risk it,
except in a perfectly sure business; will make money slowly, yet lose
little; and with large intellectual organs, in the long run acquire considerable wealth: p. 136.

MODERATE. — With large Cautiousness, make few promises; but
with large Conscientiousness, scrupulously fulfill them, because promise only what can be performed; with small Self-esteem, and large
Veneration, Conscientiousness, and Cautiousness, if a professed Chris-
tian, will have many fears about future salvation; with only average
propensities, will lack energy, enterprise, and fortitude; with large
Firmness and Cautiousness, are very slow to embark, yet once committed, rarely give up; with large reasoning faculties, may be sure
of success, because see why and how it is to be brought about; with
large Acquisitiveness, will hold on to whatever money is once ac
quired, or at least spend very cautiously, and only where sure to be
returned with interest; should cheer up, never despond, count favorable, but not unfavorable chances, keep up a lively, buoyant state of
mind, and " hope on, hope ever :" p. 139.

SMALL. — Expect and undertake very little ; with large Cautiousness, put off too long ; are always behind ; may embark in projects after everybody else has succeeded, but will then be too late, and in general knock at the door just after it has been bolted; with large Cautiousness, are forever in doubt; with large Approbativeness and Cautiousness, though most desirous of praise, have little hopes of ob taining it, and therefore exceedingly backward in society; yet fear ridicule rather than hope for praise ; are easily discouraged, see lions in the way, lack enterprise, magnify obstacles, etc.: p. 140.

VERY SMALL. — Expect next to nothing, and undertake less : p. 140.

TO CULTIVATE. — Look altogether on the bright side, the dark none ; calculate all the chances for, none against you ; mingle in young and lively society ; banish care, and cultivate juvenility ; cheer up ; venture more in business ; cultivate trust in the future, and " look aloft ! " 272.

TO RESTRAIN. — Offset excessive expectation by intellect ; say to yourself, " My hope so far exceeds realities that I shall not get half I expect," and calculate accordingly ; do business on the *cash* principle, in both buying and selling, otherwise you are in danger of buying more than you can pay for, and indorsing too much ; build no castles in the air; indulge no revelings of hope; shoulder only half the load you feel confident you can carry, and balance your visionary anticipations by cool judgment: 273.

19. SPIRITUALITY.

THE PROPHET. — Intuition ; faith ; prescience; the "light within :" trust in Providence ; prophetic guidance ; perception and feeling of the spiritual; interior perception of truth, what is best, what is about to transpire, etc.; that which foresees and warns.

Adapted to man's prophetic gift, and a future life. Perversion — superstition ; witchcraft; and with Cautiousness large, fear of ghosts.

VERY LARGE. — Are led and governed by a species of prophetic guiding ; feel by intuition what is right and best; are forewarned of danger, and led by spiritual monitions into the right way; feel internally what is true and false, right and wrong, best and not best; unless well regulated, are too credulous, superstitious, and a believer in dreams, ghosts, and wonders, and liable to be misled by them and so-called prophecies, as well as to become fanatical on religion: p. 143

LARGE. — Perceive and know things independent of the senses or intellect, or, as it were, by prophetic intuition; experience an internal consciousness of what is best, and that spiritual communion which constitutes the essence of true piety; love to meditate; experience a species of walking clairvoyance; are as it were "forewarned;" combined with large Veneration, hold intimate communion with the Deity, who is profoundly adored, and take a world of pleasure in that calm, happy, half-ecstatic state of mind caused by this faculty; with large Causality, perceive truth by intuition, which philosophical tests prove correct; with large Comparison added, have a deep and clear insight into spiritual subjects, and embody a vast amount of the highest order of truth; and clearly perceive and fully realize a spiritual state of being after death: p. 142.

FULL. — Have a full share of high, pure, spiritual feeling; many premonitions or interior warnings and guidings, which, implicitly followed, conduct to success and happiness through life; and an inner test or touchstone of truth, right, etc., in a kind of interior consciousness, which is independent of reason, yet, unperverted, in harmony with it; are quite spiritual-minded, and, as it were, "led by the spirit:" p. 143.

AVERAGE. — Have some spiritual premonitions and guidings, yet they are not always sufficiently distinct to be followed; but when followed, they lead correctly; see this "light within," and feel what is true and best with tolerable distinctness, and should cultivate this faculty by following its light: p. 141.

MODERATE. — Have some, but not very distinct perception of spiritual things; rather lack faith; believe mainly from evidence, and little from intuition; with large Causality, say "Prove it," and take no man's say without good *reasons:* p. 144.

SMALL. — Perceive spiritual truths so indistinctly as rarely to admit them; are not guided by faith, because so weak; like disbelieving Thomas, must see the fullest *proof* before believing; have very little credulity, and doubt things of a superhuman origin or nature; have no premonitions and disbelieve in them, and lack faith: p. 145.

VERY SMALL. — Have no spiritual guidings or superstitions: p. 146.

TO CULTIVATE. — Muse and meditate on divine things, the

Deity, a future existence, the state of man after death, immortality, and that class of subjects; and especially, *follow* your innermost impressions or presentiments in everything, as well as open your mind to the intuitive perception of truth : 276.

To Restrain. — Cultivate the terrestrial more and celestial less, abstain from and restrain spiritual musings and contemplations, and confine yourself more to the practical, tangible, and real; keep away from fanatical meetings, and confine yourself more to life as it is — to what and where you are, instead of are to be, to earth, its duties and pleasures : 277.

20. VENERATION.

The Worshipper. — Devotion; adoration of a Supreme Being; reverence for religion and things sacred; love of prayer, religious rites, worship, ordinances, observances, etc. Obedience; conservatism. Adapted to the existence of a God, and the pleasures and benefits man experiences in His worship. Perverted, it produces idolatry, superstitious reverence for authority, bigotry, religious intolerance, etc.

VERY LARGE. — Experience the highest degree of Divine love and worship; place God as supreme upon the throne of the soul, and make His worship a central duty; manifest extreme fervor, anxiety, and delight in divine worship, and are preëminently fervent in prayer; feel obsequious reverence for age, time-honored forms, ceremonies, and institutions; with moderate

VERY LARGE.
20

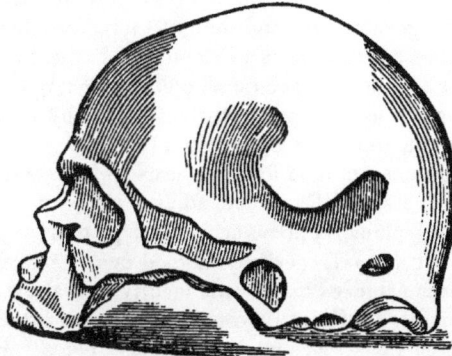

No. 164. — DIANA WATERS, who went about Philadelphia praying and exhorting all she met to repent and pray to God.

Self-esteem, and large Conscientiousness and Cautiousness, and a disordered temperament, experience the utmost unworthiness and guiltlness in His sight, and are crushed by a sense of sin and vileness, yet should never cherish these feelings; are always dreading the wrath of Heaven, no matter whether the actions are right or wrong; and should cultivate religious cheerfulness and hope of future happiness: p. 149.

VERY SMALL.

20

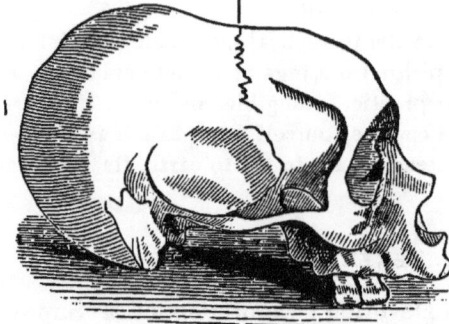

No. 165. — A NEGRO MURDERER, WHO IGNORED ALL RELIGION.

LARGE. — Experience an awe of God and of things sacred; love to adore the Supreme Being, especially in His works; feel true devotion, fervent piety, and love of divine things; take great delight in religious exercises; have much respect for superiority; regard God as the centre of hopes, fears, and aspirations; with large Hope and Spirituality, worship Him as a Spirit, and hope to be with and like Him; with large Ideality, contemplate His works with rapture and ecstasy; with large Sublimity, adore Him as infinite in everything; with large reasoning organs, have clear, and, if the faculties are evenly developed, unperverted, correct ideas of the Divine character and government, and delight to reason thereon; with large Parental Love, adore Him as a Friend and Father; and with large Benevolence, for His infinite goodness, etc.; with large Causality added, as securing the happiness of sentient beings by a wise institution of law, and as the great first CAUSE of all things; with large and perverted Cautiousness, mingle fear and dread with worship; with large Constructiveness and Causality, admire the system evinced in His architectural plans, contrivances, etc.: p. 148.

FULL. — Experience a good degree of religious worship whenever circumstances excite this faculty, but allow the stronger faculties frequently to divert it, yet pray at least internally; with large or very large Conscience or Benevolence, place religion in doing right and doing good more than in religious observances, and esteem duties higher than ceremonies; with strong propensities, may be devout upon the Sabbath, yet will be worldly through the week, and experience some conflict between the religious and worldly aspirations, etc.: p. 149.

AVERAGE. — Will adore the Deity, yet often make religion subservient to the larger faculties; with large Adhesiveness, Benevolence, and Conscience, may love religious meetings, to meet friends, and pray for the good of mankind, or because duty requires attendance; yet are not habitually and innately devotional, except when this faculty is especially excited by circumstances: p. 147.

MODERATE. — Will not be particularly devout or worshipful; with large Benevolence and Conscientiousness, if religiously educated, may be religious, yet will place religion more in works than faith, in duty than prayer, and be more moral than pious; in prayer will supplicate blessings upon mankind, and with Conscientiousness large, will confess sin more than express an awe of God; with large reflectives, worship no further than reason precedes worship; with moderate Spirituality and Conscientiousness, care little for religion as such, but with large Benevolence, place religion mainly in doing good, etc.; and are by no means conservative in religion, but take liberal views of religious subjects; and are religious only when this faculty is considerably excited: p. 150.

SMALL. — Experience little devotion or respect, and are deficient in fervor; care little for religious observances, and are not easily impressed with the worshipping sentiment: p. 150.

VERY SMALL. — Are almost destitute of the feeling and practice of this sentiment.

To CULTIVATE. — Study and admire the divine in nature, animate and inanimate, heaven and earth, man and things, present and future; cultivate admiration and adoration of the Divine character and government, of this stupendous order of things, of the beauties and perfections of nature, as well as a regard for religion and things sacred; but contemplate the Divine mercy and goodness rather than austerity, and salvation than condemnation: 279.

To RESTRAIN is rarely, if ever, necessary, unless where religious excitement engenders religious fanaticism and hallucination. In such cases avoid religious meetings, conversations, etc., as much as possible; cultivate the other faculties, and especially those which relate to his world and its pleasures; take those physical remedies, exercise, bathing, etc., which will withdraw blood from the head, and promote health; and especially do not think of the Deity with feelings of fear or terror, but as a kind and loving heavenly Father, good to all His creatures: 280.

21. BENEVOLENCE.

LARGE.
21

SMALL.

21

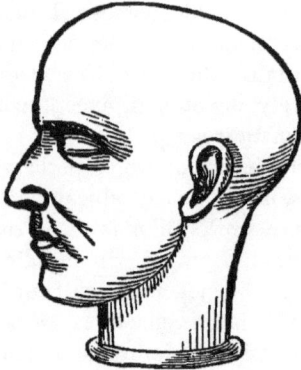

No. 166. — MR. GOSSE — GAVE AWAY
TWO FORTUNES.

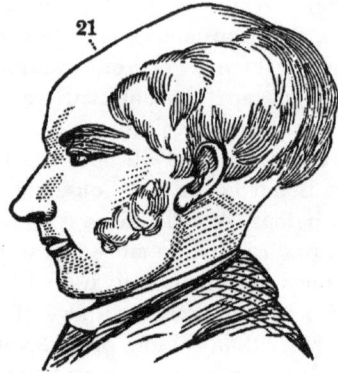

No. 167. — JUDAS, JR.

THE GOOD SAMARITAN. — Goodness; philanthropy; generosity; sympathy; kindness; humanity; desire to make others happy; a self-sacrificing disposition; the accommodating, neighborly spirit.

Adapted to man's capability of making his fellow-men happy. Perversion — misplaced sympathy, and maudlin philanthropy.

VERY LARGE. — Are deeply and thoroughly imbued with a benevolent spirit, and do good spontaneously; with large Adhesiveness and moderate Acquisitiveness, are too ready to help friends; and with large Hope added, especially inclined to indorse for them; with large Acquisitiveness, bestow time more freely than money, yet will also give the latter; but with only average or full Acquisitiveness, freely bestow both substance and personal aid; with large Veneration and only full Acquisitiveness, give freely to religious objects; with large Combativeness and Destructiveness, are more severe in word than deed, and threaten more than execute; with larger moral than animal organs, literally overflow with sympathy and practical goodness, and reluctantly cause others trouble; with large reasoning organs, are truly philanthropic, and take broad views of reformatory measures; with large Adhesiveness and Parental Love, are preëminently qualified for nursing; with large Causality, give excellent advice, etc., and should not let sympathy overrule judgment : p. 157.

LARGE. — Delight to do good; make personal sacrifices to render others happy; cannot witness pain or distress, and do what can well be done to relieve them; manifest a perpetual flow of disinterested goodness; with large Adhesiveness, Ideality, and Approbativeness, and only average propensities and Self-esteem, are remarkable for practical goodness; live more for others than self; with large domestic organs, make great sacrifices for family; with large reflectives, are perpetually reasoning on the evils of society, the way to obviate them, and to render mankind happy; with large Adhesiveness, are hospitable; with moderate Destructiveness, cannot witness pain or death, and revolt at capital punishment; with moderate Acquisitiveness, give freely to the needy, and never exact dues from the poor; with large Acquisitiveness, help others to help themselves rather than give money; with large Combativeness, Destructiveness, Self-esteem, and Firmness, at times evince harshness, yet are generally kind: p. 155.

FULL. — Show a good degree of kind, neighborly, and humane feeling, except when the selfish faculties overrule it, yet are not remarkable for disinterestedness; with large Adhesiveness, manifest kindness toward friends; with large Acquisitiveness, are benevolent when money can be made thereby; with large Conscientiousness, are more just than kind, and with large Combativeness and Destructiveness, are exacting and severe toward offenders: p. 158.

AVERAGE. — Manifest kindness only in conjunction with Adhesiveness and other large organs; and with only full Adhesiveness, if kind, are so for selfish purposes; with large Acquisitiveness, give little or nothing, yet may sometimes do favors; with large Veneration, are more devout than humane; and with only full reasoning organs, are neither philanthropic or reformatory: p. 153.

MODERATE. — Allow the selfish faculties to infringe upon the happiness of others; with large Combativeness, Destructiveness, Self-esteem, and Firmness, are comparatively hardened to suffering; and with Acquisitiveness and Secretiveness added, evince almost unmitigated selfishness.

SMALL. — Care little for the happiness of man or brute, and do still less to promote it; make no disinterested self-sacrifices; are calous to human woe; do few acts of kindness, and those grudgingly, and have unbounded selfishness: p. 159.

VERY SMALL. — Feel little and evince none of this sentiment, but are selfish in proportion as the other faculties prompt: p. 159.

TO CULTIVATE. — Be more generous and less selfish; and more

kind to all, the sick included; interest yourself in their wants and woes, as well as their relief; and cultivate general philanthropy and practical goodness in sentiment and conduct; indulge benevolence in all the little affairs of life, in every look and action, and season your whole conduct and character with this sentiment: 282.

To Restrain. — Lend and indorse only where you are willing and can afford to lose; give and do less freely than you naturally incline to; bind yourself solemnly not to indorse beyond a given sum; harden yourself against the woes and sufferings of mankind; avoid waiting much on the sick, lest you make yourself sick thereby, for your Benevolence is in danger of exceeding your strength; be selfish first and generous afterward, and put Benevolence under bonds to judgment: 283.

THE SELF-PERFECTING GROUP.

Love and talent for the fine arts; improvement, self-perfection, and obtaining and acquiring whatever is beautiful and perfect.

This group elevates and chastens the animal faculties; prevents the propensities, even when strong, from taking on the grosser sensual forms of action, and hence is rarely found in criminals; elevates even the moral sentiments, and constitutes a stepping-stone from the animal to the moral, and a connecting link between the moral and the intellectual in man.

Very Large. — Perfectly abhor the coarse, low, sensual, carnal, and animal action of the propensities, and follow after the beautiful and perfect in nature and art; with strong propensities, manifest them in a proper manner; with a large moral lobe, adopt imposing and eloquent forms of religion, as the Episcopalian, etc.

Large. — Aspire after a higher and more perfect state or style of feeling and character and conduct; revolt at the imperfect and sensual in all their forms; and are like Very Large, only less so.

Full. — Like style, but can live without it; are like Large in quality, only less in degree.

Average. — Have only commonplace aspirations after a high life, love of the fine arts, etc.

Moderate. — Are comparatively indifferent to the beauties of nature and art, fail both to appreciate and adopt them, and prefer common houses, clothes, furniture, and style of living to the artistical and stylish, and feel out of place when surrounded by the elegances of life with large Veneration, have a rude religion, etc.

SMALL. — Are rude, uncultivated, contented with few and plain articles of dress, furniture, property, etc., and prefer the rudeness of savage to the elegances of civilized life.

VERY SMALL. — Are almost destitute of these perfecting aspirations and sentiments.

To CULTIVATE. — Associate with persons of wit, ingenuity, and refinement; visit galleries of art and mechanism, scenes of beauty and perfection, and read poetry and other works of the most polished and refined writers.

To RESTRAIN. — Give more attention to the common affairs of life, and refrain from fostering æsthetic subjects; read history, science, and metaphysics, rather than poetry, romance, etc.

22. CONSTRUCTIVENESS.

THE MECHANIC. — Manual skill; ingenuity; the making instinct; the tool-using talent; sleight of hand in constructing things; invention; love of machinery.

Adapted to man's need of things made, such as houses, clothes, and manufactured articles·of all kinds. Perverted, it wastes time and money on perpetual motion, and other like futile contrivances.

VERY LARGE. — Show extraordinary mechanical ingenuity, and a perfect passion for making everything; with large Imitation, Form, Size, and Locality, have first-rate talents as an artist, and for drawing, engraving, etc.; and with Color added, are excellent limners; with Ideality, add manual skill; with large Causality and perceptives, add invention to execution, etc.: p. 162.

LARGE. — Love to make, are able and disposed to tinker, mend, and fix up, build, manufacture, employ machinery, etc.; show mechanical skill and dexterity in whatever is done with the hands; with large Causality and perceptives, are inventive; and with large Imitation added, can make after a pattern, and both copy the improvements of others, and supply defects by original inventions, as well as improve on the mechanical contrivances of others; make heal save hands of self and others; are a natural boss, and direct

VERY LARGE. SMALL.

No. 168. — JACOB JORDAN.　　No. 169. — LORD LIVERPOOL.

work and working men to excellent advantage; with the mental temperament, and large intellectual organs and Ideality, employ ingenuity
in constructing sentences and arranging words, and forming essays,
sentiments, books, etc. : p. 161.

FULL. — Can, when occasion requires, employ tools and use the
hands in making, tinkering, and fixing up, and turn off work with
skill, yet have no great natural passion or ability therein ; with practice, can be a good workman ; without it, would not excel : p. 163.

AVERAGE. — Like full, only less gifted in this respect : p. 160.

MODERATE. — Are rather awkward in the use of tools, and in
manual operations of every kind; with large Causality and perceptives, show more talent to invent than execute, yet little in either ;
with the mental temperament, evince some mental construction, yet
not much manual ingenuity : p. 163.

SMALL. — Are deficient in the tool-using capability ; awkward in
making and fixing up things ; poor in understanding and managing
machinery ; take hold of work awkwardly and wrong end first ; write
poorly, and lack both kinds of construction : p. 163.

VERY SMALL. — Can make nothing, and are most awkward.

To CULTIVATE. — Try your hand in using tools, and turning off
work of any and every kind ; if in any writing business, try to write
well and cut flourishes ; if a mechanic, do with skill and dexterity
what you undertake, etc. ; observe and study machinery and inventions, and call out this faculty in its various phases by *work:* 285.

To Restrain. — Give yourself more to the exercise of your other faculties, and less to mechanical inventions and executions; especially attempt no chimerical inventions, perpetual motion, and the like ; and spend no more time or money on them than you can spare without inconvenience : 286.

23. IDEALITY.

VERY LARGE.

No. 170. — CLYTIE.

The Poet. — Taste ; refinement : imagination ; perception and admiration of the beautiful and perfect ; purity of feeling ; sense of propriety ; polish ; love of perfection, purity, poetry, flowers, beauty, elegance, gentility, the fine arts, etc. ; personal neatness ; finish.

Adapted to the beautiful in nature and art. Perverted, it gives fastidiousness and extra niceness.

Very Large. — Have the highest order of taste and refinement; love the exquisite and perfect beyond expression, and are correspondingly dissatisfied with the imperfect, especially in themselves ; admire beauty in bird and insect, flower and fruit, animal and man, the physical and mental ; are perfectly enraptured with the impassioned, oratorical, and poetical in speech and action, in nature and art, and live much in an ideal world ; have a most glowing and vivid imagination, and give a delicate finish to every act and word, thought and feeling, and find few things to come up to their exalted standard of taste ; with only average Causality, have more taste than solidity of mind and character, and more exquisiteness than sense ; but with large reflectives, add the highest artistical style of expression to the highest conceptions of reason, and with organic quality 6 or 7, are always and involuntarily eloquent.

Large. — Appreciate and enjoy beauty and perfection wherever found, especially in nature; give grace, purity, and propriety to expression and conduct, gracefulness and polish to manners, and general good taste to all said and done; are pure-minded ; enjoy the ideal of poetry and romance; desire to perfect character, and obviate blemishes, and with Conscientiousness large, moral imperfections; with large social organs, evince a nice sense of propriety in friendly intercourse; eat in a becoming and genteel manner; with large moral organs, appreciate perfection of character, or moral beauties and excellences most; with large reflectives, add a high order of sense and strength of mind to beauty and perfection of character; with large perceptives, are gifted with a talent for the study of nature, etc.: p. 166.

FULL. — Evince a good share of taste and refinement, yet not a high order of them, except in those things in which it has been vigorously cultivated; with large Language, Eventuality, and Comparison, may compose with elegance, and speak with some eloquence, yet will have more force of thought than beauty of diction; with large Constructiveness, will use tools with fair taste, yet more skill; with large Combativeness and Destructiveness, show general refinement, except when provoked, but are then grating and harsh; with large moral organs, evince more moral beauty and harmony than personal neatness; with large intellectual organs, possess more beauty of mind than regard for looks and outside appearances, and prefer the sensible to the elegant and nice, etc.: p. 168.

AVERAGE. — Prefer the plain and substantial to the ornamental, and are utilitarian; with large intellectual organs, prefer sound, solid matter to the ornaments of style, and appreciate logic more than eloquence; with Benevolence and Adhesiveness large, are hospitable, and evince true cordiality, yet care nothing for ceremony; with Approbativeness large, may try to be polite, but make an awkward attempt, and are rather deficient in taste and elegance; with Constructiveness large, make things that are solid and serviceable, but do not polish them off; with Language large, talk directly to the purpose, without paying much attention to expression, etc.: p. 160.

MODERATE. — Rather lack taste in manners and expression; have but little of the sentimental or finished; should cultivate harmony and perfection of character, and endeavor to polish up; with strong propensities, evince them in rather a coarse and gross manner, and are more liable to their perverted action than when this organ is large, and are homespun in everything: p. 163.

SMALL. — Show a marked deficiency in whatever appertains to taste and style, also to beauty and sentiment: p. 163.

VERY SMALL. — Are almost deficient in taste, and evince none

To Cultivate. — First avoid all disgusting habits, like swearing chewing, drinking, low conversation, vulgar expressions, and associates, etc.: dress and appear in good taste, and cultivate personal neatness, good behavior, refinement and style in manners, purity in feeling, the poetical and sentimental, and elegant and classical style of conversation, expression, and writing, and love of the fine arts and beautiful forms; of the beauties of nature, of sunrise, sunset, mountain, lawn, river, scenery, beautiful birds, fruits, flowers, mechanical fabrics, and productions, — in short, the beautiful and perfect in nature, in general, and yourself in particular : 288.

To Restrain. — Remember that in you the ideal and imaginative exceed the practical; that your building airy castles out of bubbles prevents your building substantial structures, and attaining useful life ends ; that you are too symbolical, fastidious, and ornamental, too much tormented by spots and wrinkles, too apt to discard things that are almost perfect, because not *quite* so, and hold in check the revelings of Ideality, and learn to prize what is right, instead of discarding the greater good because of minor faults. Especially do not refuse to associate with others because they are not in all particulars just to your fastidious tastes : 289.

24. SUBLIMITY.

Infinitude. — Perception and love of the grand, infinite, vast, illimitable, omnipotent, eternal, and endless.

Adapted to that infinitude which characterizes every department of nature. Perverted, it leads to bombast, and a wrong application of extravagant words and ideas.

Very Large. — Have a literal passion for wild mountain scenery, and the towering, romantic, boundless, endless, infinite, eternal. and stupendous, and are like Large, only more so.

Large. — Appreciate and admire the grand, sublime, vast, and magnificent in nature and art, and enjoy exceedingly mountain scenery, thunder, lightning, tempests, vast prospects, and all that is awful and magnificent; also the foaming, dashing cataract, a storm at sea, the lightning's vivid flash, and its accompanying thunder ; the commotion of the elements, and the star-spangled canopy of heaven, and all manifestations of omnipotence and infinitude; with large Veneration, are particularly delighted by the infinite as appertaining to the

Deity, and His attributes and works; and with large Time added, have unspeakably grand conceptions of infinitude as applicable to eternity, past and future; with large intellectual organs, take a comprehensive view of subjects, and give illimitable scope to all mental investigations and conceptions, so that they can be carried out to any extent; and with Ideality large, add the beautiful and perfect to the sublime and infinite.

FULL. — Enjoy grandeur, sublimity, and infinitude quite well, and impart considerable of this element to thoughts, emotions, and expressions, and evince the same qualities as Large, only in a less degree.

AVERAGE. — Possess considerable of this element, when it is powerfully excited, yet, under ordinary circumstances, manifest only an ordinary share of it.

MODERATE. — Are rather deficient in conception and appreciation of the illimitable and infinite; and with Veneration moderate, fail to appreciate this element in nature and her Author.

SMALL. — Show a marked deficiency in this respect, and should earnestly cultivate it.

VERY SMALL. — Are almost destitute of sublime emotions.

TO CULTIVATE. — Mount the lofty summit and contemplate the outstretched landscape; admire the grand and stupendous in towering mountain, rolling cloud, rushing wind and storm, loud thunder, majestic river, raging sea, roaring cataract, burning volcano, and the boundless, infinite, and eternal in nature and her Author: 291.

TO RESTRAIN — which is rarely ever necessary — refrain from the contemplation of the sublime: 292.

25. IMITATION.

THE MIMIC. — Conformity; ability and desire to copy, take pattern, imitate, do and become like, mock, etc.

Adapted to man's requisition for doing, talking, acting, etc., like others. Perverted, it copies even faults.

VERY LARGE. — Can mimic, act out, and pattern after almost anything; with large Mirthfulness, relate anecdotes to the very life; have a theatrical taste and talent; gesticulate almost constantly while speaking; and with large Language, impart an uncommon amount of *expression* to countenance, and everything said; with large Individuality, Eventuality, Language, Comparison, and Ideality, can make a splendid speaker; and with large Mirthfulness, and full Secretiveness added, can keep others in a roar of laughter, yet remain serious; with an uneven head, are droll and humorous in the extreme; with large Approbativeness, delight in being the sport-maker

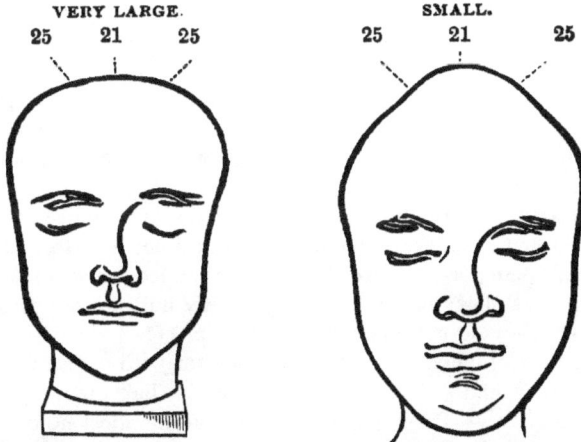

No. 171. — Clara Fisher.   No. 172. — Jacob Jervis.

at parties, etc., and excel therein; with large Constructiveness, Form, Size, Locality, and Comparison, full Color, and a good temperament, and a full-sized brain, can make a very superior artist of almost any kind; but with Color small, can engrave, draw, carve, model, etc., better than paint: p. 171.

Large. — Have a great propensity and ability to copy and take pattern from others, and do what is seen done; describe and act out well; with large Language, gesticulate much; with large perceptives, require to be shown but once; with large Constructiveness, easily learn to use tools, and to make things as others make them; and with small Continuity added, are a jack-at-all-trades, but thorough in none; begin many things, but fail to finish; with large Causality, perceptives, and an active temperament added, may make inventions, or improvements, but never dwell on one till it is complete, or are always adding to them; with large Approbativeness, copy after renowned men; with large Adhesiveness, take pattern from friends; with large Language, imitate the style and mode of expression of others; with large Mirthfulness and full Secretiveness, create laughter by taking off the oddities of people; with large Form, Size, and

Constructiveness, copy shape and proportions; with large Color, imi tate colors, and thus of all the other faculties : p. 170.

FULL. — Copy quite well, yet not remarkably so; with large Caus‑ality, had rather invent a new way of doing things than copy the ordi‑nary mode, and evince considerable imitating talent when this faculty works with large organs, yet but little otherwise: p. 171.

AVERAGE. — Can copy tolerably well when this faculty is strongly excited, yet are not a natural mimic, nor a copyist; with only full Constructiveness, evince little manual dexterity ; yet with large Caus‑ality, can originate quite well, and show no great disposition or abil‑ity to copy either the excellences or deficiencies of others, but prefer to be original : p. 169.

MODERATE. — Have little inclination to do what and as others do ; but with large Causality, prefer to strike out a new course, and invent an original plan ; with large Self-esteem added, have an excellent conceit of that plan ; but if Causality is only fair, are full of original devices, yet they do not amount to much : p. 171.

SMALL. — Copy even commonplace matter with extreme difficulty and reluctance, and generally do everything in their own way : p. 172.

VERY SMALL. — Possess scarcely any, and manifest no disposition or ability to copy anything, not even enough to learn to talk well : p. 173.

To CULTIVATE. — Practice copying from others in manners, ex‑pressions, sentiments, ideas, opinions, everything, and try your hand at drawing, and in every species of copying and imitation as well as conforming to those around you; that is, try to become what they are, and do what and as they do: 294.

To RESTRAIN. — Maintain more your own personality in thought, doctrine, character, everything, and be less a parrot and echo, and cultivate the original and inventive in everything: 295.

26. MIRTHFULNESS.

THE LAUGHER. — Wit; facetiousness; ridicule; love of fun; disposition and ability to joke, and laugh at what is ill‑timed, improper, or unbecoming; laughter; intuitive percep‑tion of the ridiculous; pleasantness; facetiousness

Adapted to the absurd, inconsistent, and laughable. Per‑verted, it makes fun on solemn occasions, and where there is nothing ridiculous at which to laugh.

VERY LARGE. — Show an extraordinary disposition and capacity to make fun; are always laughing and making others laugh; with large Language, Comparison, Imitation, perceptives, and Adhesiveness, and moderate Self-esteem and Secretiveness, are " the fiddle of the company; " with only average Ideality added, are clownish, and often say undignified, and perhaps low things, to raise a laugh; and with only moderate Causality, things that lack sense, etc. : p. 175.

LARGE. — Enjoy a hearty laugh at the absurdities of others exceedingly, and delight to make fun out of everything not exactly proper or in good taste, and are always ready to give as good a joke as get; with large Amativeness, love to joke with and about the other

VERY LARGE.   SMALL.

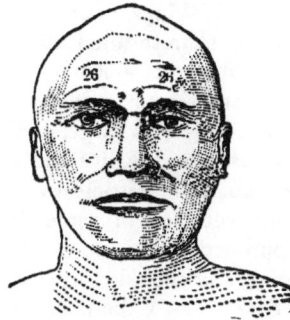

No. 173. — LAURENCE STERNE.   No. 174. — INDIAN CHIEF.

sex; and with large Imitation and Language added, to talk with and tell stories to and about them; with large Combativeness and Ideality added, make fun of their imperfections in dress, expression, manners, etc., and hit them off to admiration; with large Adhesiveness, Language, and Imitation, are excellent company; with large Causality, Comparison, and Combativeness, argue mainly by ridicule or by showing up the absurdity of the opposite side, and excel more in exposing the fallacy of other systems than in propounding new ones; with large Ideality show taste and propriety in witticisms, but with this faculty average or less, are often gross, and with large Amativeness added, vulgar in jokes; with large Combativeness and Destructiveness, love to tease, and are sarcastic, and make many enemies; and with large

Comparison added, compare those disliked to something mean, disgusting, and ridiculous : p. 173.

FULL. — Possess and evince considerable of the fun-making disposition, especially in the direction of the larger organs; with large or
very large Comparison, Imitation, and Approbativeness, and moderate Self-esteem, manifest more of the laughable and witty than is
really possessed; may make much fun and be called witty, yet it will
be owing more to what may be called drollery than pure wit; with
moderate Secretiveness and Self-esteem, and an excitable temperament, let fly witty conceptions on the spur of the moment, and thus
increase their laughableness by their being well-timed, unexpected,
sudden, etc. : p. 175.

AVERAGE. — Are generally serious and sedate, except when this
faculty is excited, yet then often laugh heartily, and evince considerable wit; with large Individuality and Language, often say many
laughable things, yet owe wit more to argument or the criticism they
embody than to this faculty : p 172.

MODERATE. — Are generally serious, sedate, and sober, and with
large Self-esteem, stern and dignified, nor companionable except when
Adhesiveness is large, and in company with intimate friends; with
only average Ideality and Imitation, are very poor in joking, have to
explain their witticisms, and thereby spoil them; have some witty
ideas, yet lack in perceiving and expressing them; and with large
Approbativeness and Combativeness, are liable to become angry when
joked, and should cultivate this faculty by making more fun : p. 176.

SMALL. — Make little fun ; are slow to perceive, and still slower
to turn jokes; seldom laugh, and think it foolish or wrong to do so
with only average Adhesiveness, are uncompanionable ; with large reflectives and Language, may write well yet debate poorly : p. 177.

VERY SMALL. — Have few, if any, witty ideas and conceptions.

TO CULTIVATE. — Rid yourself of the idea that it is sinful or undignified to laugh ; try to perceive the witty and facetious aspects
of subjects and things; cultivate the acquaintance of mirthful people,
and read witty books, and try to imbibe their spirit : 297.

TO RESTRAIN. — Cease hunting for something to laugh at and
make fun of; observe in the conduct and appearance of others all
that is congruous, correct, and proper, and not that merely which is
droll or ridiculous ; avoid turning everything into ridicule, punning,
playing upon words, double entendre, etc. : 298.

INTELLECTUAL FACULTIES.

Knowing, remembering, and reasoning powers; intellectual capability.

Adapted to the physical and metaphysical. Perverted, they apply their respective powers to accomplish wrong ends.

VERY LARGE.—Have natural greatness of intellect and judgment, and a high order of talents and sound sense, with originality, capaciousness, and comprehensiveness of mind which can hardly fail to make their mark.

LARGE.—Possess sufficient natural talent and intellectual capability to take a high stand among men; and have strength of mind, superior judgment, and power both of acquiring knowledge easily and reasoning profoundly. Their direction depends upon the other faculties; with large animal organs and weak morals, they make philosophical sensualists; with large moral and weaker animal organs, moral and religious philosophers, etc.

FULL.—Have good intellectual capabilities, and much strength of mind, provided it is well cultivated; with large Acquisitiveness, a talent to acquire property; with large moral organs, to enlighten and improve the moral character; with large Constructiveness, mechanical intelligence, etc.

AVERAGE.—Evince fair mental powers, provided they are cultivated, otherwise only moderate; with an excitable temperament, allow the feelings and stronger faculties to control judgment; with large moral organs, have more piety than talents, and allow religious prejudices and preconceived doctrines to prevent impartial intellectual examination; with moderate Acquisitiveness, will never acquire property; with average Constructiveness, will be a poor mechanic, etc.

MODERATE.—Are rather deficient in sense and judgment, yet not palpably so; can be easily imposed upon; lack memory, and are rather wanting in judgment, comprehension, and intellectual capacity.

SMALL.—Are decidedly deficient in mind; slow and dull of comprehension; lack sense, and have poor powers of memory and reason.

VERY SMALL.—Are naturally idiotic.

These faculties are divided into three classes—the Perceptive, the Literary, and the Reflective—which, when large, confer these three kinds of talent, practicality, scholarship, and originality.

To CULTIVATE.—Exercise the whole mind in diversified studies and intellectual exercises. See specific directions in "Fowler on

Memory." And probably nothing is as well calculated to discipline and improve intellect as the study and practice of Phrenology.

To RESTRAIN. — Divert the flow of blood from the brain to the body by vigorous exercise, an occasional hot bath, frequent ablutions, and a general abstinence from intellectual exercises, especially reading and writing.

THE PERCEPTIVE FACULTIES.

These bring man into direct intercourse with the physical world; take cognizance of the physical qualities of material things; give correct judgment of the material properties of things, and a practical cast of mind.

VERY LARGE. — Are preëminent in these respects; know by intuition the conditions, fitness, value, etc., of things; have extraordinary power of observation, and ability to acquire knowledge, and a natural tas* : for examining, collecting statistics, studying the natural sciences, et . For combinations, see Large.

LARGE. — Judge correctly of the various qualities and relations of material things; with Acquisitiveness large, form correct ideas of the value of property, goods, etc., and what kinds are likely to rise in value, and make good bargains; with large Constructiveness, can conduct mechanical operations, and have very good talents for building machinery, superintending workmen, etc.; with the mental temperament, and large reflectives added, are endowed by nature with a truly scientific cast of mind, and a talent for studying the natural sciences, and are useful in almost every department and situation in life; with an active temperament and favorable opportunities, know a good deal about matters and things in general; are quick of observation, and perception, and matter-of-fact, common-sense tact, and will show off to excellent advantage; appear to know all; have superior talents for acquiring and retaining knowledge with facility, and attending to the details of business, becoming an excellent scholar, etc.; and have a strong thirst after knowledge.

FULL. — Have fair perceptive powers, and a good share of practical sense; learn and remember most things quite well; love reading and knowledge, and with study can become a good scholar, yet not without it; with large Acquisitiveness, judge of the value of things with sufficient correctness to make good bargains, but with moderate Acquisitiveness lack such judgment; with large Constructiveness,

aided by experience, have a good mechanical mind, but without experience, or with only moderate Constructiveness, are deficient.

AVERAGE. — Are endowed with only fair perceptive and knowing powers, but, well cultivated, know considerable about matters and things, and learn with tolerable ease; yet without cultivation are deficient in practicability of talent, and capability of gathering and retaining knowledge. For combinations, see Full.

MODERATE. — Are rather slow and dull of observation and perception, require some time to understand things, and even then lack specific knowledge of detail; are rather deficient in matter-of-fact knowledge, and show off to poor advantage; learn slowly, and fail in off-hand judgment and action; with only average Acquisitiveness, are deficient in judging of the value of things, and easily cheated; and with moderate Language, are rather wanting in practical talent, and cannot show advantageously what is possessed.

SMALL. — Are very deficient in recollecting and judging; lack practical sense, and should cultivate the knowing and remembering faculties.

VERY SMALL. — See few things, and know almost nothing about the external world, its qualities, and relations.

TO CULTIVATE. — *Exercise* each separately, and all together, in examining closely all the material properties of physical bodies; study the natural sciences, especially Phrenology; examine the natural qualities of all natural objects: 403.

27. INDIVIDUALITY.

THE OBSERVER. — Cognizance of individual objects; desire to see and examine; minuteness; scrutiny; looking.

Adapted to individual existence, or the thingness of things. It is the door through which most forms of knowledge enter the mind. Perverted, it stares and gazes impudently.

VERY LARGE. — Have an insatiable desire to see and know all about everything, together with extraordinary powers of observation; cannot rest satisfied till all is known; individualize everything, and are very minute and particular in observing all things; with large Ideality, employ many allegorical and like figures; with large Human Nature and Comparison. observe every little thing which people say and do, and read character correctly from what smaller Individuality would not notice: p. 185.

LARGE. SMALL.

No. 175. — EPHRAIM BYRAM.   No. 176. — DEACON SETH TERRY

LARGE. — Have a great desire to see, know, examine, experience, etc.; are a great and practical observer of men and things; see whatever is transpiring around, what should be done, etc.; are quick of perception, knowing, and with large Acquisitiveness, sharp to perceive whatever appertains to property; with large Parental Love, whatever concerns children; with large Alimentiveness, whatever belongs to the flavor or qualities of food, and know what things are good by looking at them; with large Approbativeness or Self-esteem, see quickly whatever appertains to individual character, and whether it is favorable or unfavorable; with large Conscientiousness, perceive readily the moral, or right and wrong of things; with large Veneration, " See God in clouds, and hear him in the winds; " with large Ideality, are quick to perceive beauty, perfection, and deformity; with large Form, notice the countenances and looks of all met; with small Color, fail to observe tints, hues, and shades; with large Order and moderate Ideality, perceive disarrangement at once, yet fail to notice the want of taste or niceness. These and kindred combinations show why some persons are very quick to notice some things, but slow to observe others : p. 184.

FULL. — Have good observing powers, and much desire to see and know things, yet are not remarkable in these respects; with large Acquisitiveness, but moderate Ideality, are quick to notice whatever appertains to property, yet fail to observe instances of beauty and deformity; but with large Ideality and moderate Acquisitiveness, quickly see beauty and deformity, yet do not quickly observe the qualities of things or value of property ; with large Parental Love and Ideality, see at once indices of beauty and perfection in children ; but if Ideality and Language are moderate, fail to perceive beauty of expression or sentiment, etc. : p. 185.

AVERAGE. — Observe only the more conspicuous objects, and more in general than detail, and what especially interests : p. 183.

MODERATE. — Are rather deficient in observing disposition and capability, and should cultivate this faculty ; with large Locality, may observe places sufficiently to find them again ; with large Order, observe when things are out of place ; with large Causality, see that it may find materials for reasoning, etc. : p. 185.

SMALL. — Observe only what is thrust upon the attention, and are quite deficient in this respect : p. 186.

TO CULTIVATE. — Notice whatever comes within the range of your vision ; observe attentively all the little things done and said by everybody, all their minor manifestations of character — in short, keep a sharp lookout : 422.

TO RESTRAIN. — Look and stare less, and think more.

28. FORM.

THE DRAFTSMAN. — Configuration ; cognizance and memory of forms, shapes, faces, countenances, and looks ; perception of family likenesses, resemblances, etc.

Adapted to shape. Perverted, sees imaginary shapes of persons, things, etc., as in delirium tremens.

VERY LARGE. — Possess this capability to an extraordinary degree ; recognize persons not seen for many years ; with large Ideality, delight in beautiful forms ; with large Spirituality, see the spirits of the departed ; with disordered nerves, see horrid images, etc. : p. 188.

LARGE. — Notice, and for a long time remember, the faces, countenances, forms, looks, etc., of persons, beasts, and things once seen ; know by sight many whose name is not remembered ; with Individuality large, both observe and recollect persons and things, but with Individuality moderate, fail to notice, and hence to remember them, unless business or something special draws attention to them ; with large Parental Love, notice and recollect children, favorite animals, etc. ; with large Acquisitiveness, Individuality, and Locality, readily detect counterfeits, etc. : p. 187.

VERY LARGE. VERY LARGE.

No. 177. — RUBENS.

FORM, SIZE, AND COLOR.

No. 178.

FORM, SIZE, AND COLOR.

FULL. — Have a good recollection of the countenances of persons and shapes of things, yet not remarkably good unless this faculty has been quickened by practice, or invigorated by some strong incentive to action; with large Ideality, will recollect beautiful shapes; with large Locality and Sublimity, beautiful and magnificent scenery, etc.; and should impress the recollection of shape upon the mind: p. 188.

AVERAGE. — Have only a fair natural recollection of shapes, countenances, etc.; yet with practice may do tolerably well, but without it will be only fair in these respects, and should cultivate it: p. 186.

MODERATE. —Are rather deficient in recognizing persons and things seen; fail to recognize by their looks those who are related to each other by blood, and should cultivate this faculty by trying to remember persons and things: p. 189.

SMALL. — Have a poor recollection of persons, looks, etc.; often meet persons the next day after an introduction, or an evening interview, without knowing them; with Eventuality large, may remember their history, but not their faces; with Locality large, where they were seen, but not their looks, etc.: p. 189.

VERY SMALL. — Manifest scarcely any of this faculty: p. 189.

TO CULTIVATE. — Scan the shape of everything you would remember; study botany, conchology, Phrenology, and especially those studies which involve configuration; when talking to persons, scan eyes nose, mouth, chin, forehead, looks, expression of countenance, especially of eye, as if you were determined ever afterward to remember them ; looking at them critically, as a police detective looks at a rogue, as if saying to himself, " I'll know you, next time : " 437.

· 29. SIZE.

THE ARCHITECT. — Measurement by eye ; cognizance and memory of magnitude, quantity, bulk, distance, proportion, weight by size, etc.

Adapted to the absolute and relative magnitude of things. Perverted, it is pained by disproportion and architectural inaccuracies.

VERY LARGE. — Are endowed with an extraordinarily accurate architectural eye ; detect at one glance any departure from perfect accuracy and proportion ; often perceive errors in the work of good workmen ; can tell how high, wide, long, far, much, heavy, etc., with perfect accuracy ; judge correctly, as if by intuition, the texture, fineness, coarseness, qualities, etc., of goods ; excel in judging of property where bulk and value are to be estimated by eye ; with Constructiveness, can fit nice machinery, and in many things dispense with measuring instruments because accurate enough without, and do best on work requiring the utmost perfect accuracy : p. 191.

LARGE. — Have an excellent eye for measuring angles, proportions, disproportions, and departures therefrom, and with large Constructiveness, a good mechanical eye, and judge correctly of quantity in general ; love harmony of proportion, and are pained by disproportion ; necessary to artisans, mechanics, etc. : p. 190.

FULL. — Possess a good share of this eye-measuring power, yet are not remarkable ; with practice, do well ; without it, only fairly, and in this respect succeed well in their accustomed business : p. 191.

AVERAGE. — Have a fair eye for judging of bulk, distances, weight by size, etc., and with practice do tolerably well in it : p. 190.

MODERATE. — Measure by eye rather inaccurately, and have poor judgment of bulk, quantity, distance, etc.: p. 191.

SMALL. — Are obliged always to rely on actual measurements, because the eye is too imperfect to be trusted : p. 191.

VERY SMALL. — Are almost destitute of this faculty : p. 192.

To CULTIVATE. — Pass judgment on whatever involves how much, how heavy, how far, the centre, the amount. architectural accuracy, guessing the weight, the quantity of groceries, of everything by eye, judging how much grain to the acre, and everything involving the exercise of this faculty : p. 441.

TO RESTRAIN. — Do not allow architectural inaccuracies or any disproportion to disturb you as much as it naturally does — that is, put up with things not regulated by size and proportion.

30. WEIGHT.

THE CLIMBER. — Balancing capacity; marksmanship; intu-itive perception and application of the laws of gravity, motion, etc.; ability to balance in walking, riding, climbing aloft, etc.

Perverted, it runs imminent risk of falling by venturing too far. See illustration of Weight Large in Brunell, cut No. 184.

VERY LARGE. — Have control over the muscular system, hence can climb or walk anywhere with safety; cannot be thrown by frac-tious horses; are sure-footed; never slip or fall; are a dead shot, even "on the wing;" have an intuitive gift for skating, swimming, balancing, circus-acting, hurling, riding velocipedes, everything requir-ing muscular control; are an excellent judge of perpendiculars and levels; can plumb anything by the eye; as a sculptor or other artist, always make the picture or statue in an easy, natural, and well-bal-anced attitude, and are annoyed if the mirror, pictures, etc., do not hang plumb; with Constructiveness large, will succeed in any mechan-ical avocation requiring a steady hand, as in surgery, dental opera-tions, sleight-of-hand performances, fancy glass-blowing, etc. : p. 194.

LARGE. — Have an excellent faculty for preserving and regaining balance, riding a fractious horse, skating, carrying a steady hand, etc.; easily keep from falling when aloft or in dangerous places; are rarely seasick naturally; throw a stone, ball, or arrow straight; are pained at seeing things out of plumb; judge of perpendiculars very exactly; love to climb, walk on the edge of a precipice, etc.; with Form and Size large, are an excellent marksman; with Constructive-ness large, understand and work machinery; with Approbativeness arge, are venturesome, etc., to show what risks can be run without falling: p. 193.

FULL. — Have a good degree of this faculty, and with practice ex-cel, yet without it are not remarkable : p. 194.

AVERAGE. — Like Full, only less gifted in this respect; with only average Constructiveness and perceptives, should never engage in working machinery, because deficient in this talent: p. 192.

MODERATE.— Can keep the balance under ordinary circumstances, yet have rather imperfect control over the muscles in riding a fractious horse, or walking a narrow beam aloft, hurling, etc.; with large Cautiousness, are timid in dangerous places, and dare not venture far; are rather poor in shooting, skating, throwing, etc., unless rendered so by practice, and should cultivate this faculty by climbing, balancing, hurling, etc.: p. 194.

SMALL. — Are quite liable to seasickness, dizziness when aloft, etc.; with large Cautiousness, are afraid to walk over water, even on a wide plank, and where there is no danger; never feel safe while climbing, and fall easily: p. 195.

VERY SMALL.— Can hardly stand erect, and have very little control over the muscles: p. 195.

To CULTIVATE. — Skate, slide down hill, practice gymnastic feats, balance a long pole on your hand, walk a fence, climb, ride on horseback and velocipede, go to sea, practice gunnery, archery, throwing stones, pitching quoits — anything to exercise this faculty: 446.

To RESTRAIN.— Do not allow yourself to climb aloft, and walk narrow, dangerous places, as much as naturally inclined to. Persons often lose their lives by ambitiously attempting extraordinary feats.

31. COLOR.

THE PAINTER. — Perception, recollection, and application of colors, and delight in them.

Adapted to that infinite variety of coloring interspersed throughout Nature. Perverted, are over-particular to have colors just right.

VERY LARGE. — Have a natural taste and talent, as well as a perfect passion, for whatever appertains to colors; can carry colors perfectly in the eye, and match them from memory; take the utmost delight in viewing harmonious colors, and with very large Constructiveness, Imitation, Form, and Size, and large Weight, a full or large-sized brain, and organic quality 6 or 7, have a natural taste and talent for painting, and are a real genius in this line. For combinations, see Large.

LARGE. — Can discern and match colors by the eye with accuracy with Comparison large, can compare them closely, and detect similarities and differences; with Constructiveness, Form, Size, and Imitation large or very large, can excel in painting; but with Form and Size only average, can paint better than draw; with Ideality large, are exceedingly delighted with fine paintings, and disgusted with imperfect coloring; with large Form and Size, manage the perspective and lights and shades of painting admirably: p. 195.

FULL. — Possess a good share of coloring ability and talent, provided it has been cultivated; take much pleasure in beautiful flowers, variegated landscapes, beautifully colored fruits, etc.: p. 196.

AVERAGE. — Possess a fair share of this talent, yet are not extraordinary : p. 195.

MODERATE. — With practice, may judge of colors with considerable accuracy, yet without it will be deficient in this respect; with large Form, Size, Constructiveness, Ideality, and Imitation, may take an excellent likeness, yet will fail in the coloring: p. 197.

SMALL. — May tell primitive colors, yet rarely notice the colors of dresses, eyes, hair, etc. ; cannot describe persons and things by them, and evince a marked deficiency in coloring, taste, and talent: p. 197.

VERY SMALL. — Can hardly tell one color from another: p. 197.

TO CULTIVATE. — Observe color in general, and its shadings in particular; try to appreciate their beauties, and enjoy their richness, as seen in flower, bird, fruit, lawn, twilight, everywhere, and cultivate an appreciation of fine paintings : 450.

TO RESTRAIN is rarely necessary; go less into raptures over a new flower or painting, but give more attention to other things.

32. ORDER.

THE REGULATOR. — Method, system, arrangement; having places for things, and everything in its place ; observing business rules.

Adapted to Heaven's first law. Perverted, it overworks, annoys others to keep things in order, and is pained by disorder.

VERY LARGE. — Methodize everything; are law-abiding; governed by rules; perfectly systematic, and very particular about order, even to old-maidishness; work far beyond strength to have things just so; and with large Ideality, and an active temperament, and only fair Vitality, are liable to break down health and constitution by overworking in order to have things extra nice, and take more pains to keep things in order than this order is worth; are more nice than wise, and fastidious about personal appearance, as well as extra particular to have every little thing very nice; and with Acquisitiveness added, cannot bear to have garments soiled, and are pained in the extreme by grease-spots, ink-blots, and like deformities: p. 199.

LARGE. — Conduct business on methodical principles, and are systematic in everything; with large Acquisitiveness and Causality, have good business talents; with large Locality, have a place for everything, and everything in its place; with large Time, have a time for everything, and everything in season; with large Continuity, Comparison, and the mental temperament, have every idea, paragraph, and head of a subject in its proper place; with large Constructiveness, put and keep tools always in place, so that they can be found in the dark; with large Combativeness, are excessively vexed by disarrangement; with large Language, place every word exactly right in the sentence; with large Approbativeness, conform to established usages; with large Size, must have everything in rows, at proper distances, straight, etc.; and with large Ideality, must have everything neat and nice as well as methodical, etc.: p. 199.

FULL. — If educated to business habits, evince a good degree of method, and disposition to systematize, but without practice may sometimes show laxity; with a powerful mentality, but weaker muscles, may like to have things in order, yet do not always keep them so; with large Causality added, show more mental than physical order; with large moral organs, like to have religious matters, codes of discipline, etc., rigidly observed, and have more moral than personal method; with Acquisitiveness and perceptives large, are methodical enough for all practical purposes, yet not extra particular: p. 200.

AVERAGE. — Like order, yet may not always keep it, and desire more than practically secure: p. 198.

MODERATE. — Often leave things where they were last used, and lack method; with Ideality moderate, lack personal neatness, and should cultivate this desirable element by being more particular, but with large Ideality are more neat than systematic.

SMALL. — Have a very careless, inaccurate way of doing everything; leave things just where it happens; can never find what is wanted; take a long time to get ready, or else go unprepared, and have everything in perpetual confusion: p. 201.

VERY SMALL. — Almost wholly lack arrangement: p. 201.

13

To CULTIVATE. — Methodize and arrange everything; be regular in all your habits; cultivate system in business; have a place for everything, and keep everything in place, so that you could find it in the dark — in short, EXERCISE order: 456.

To RESTRAIN. — Work and worry less to keep order, for it costs more to keep it than it is worth. You waste your very life and strength in little niceties of order which, after all, amount to little, but are costing you your sweetness of temper and very life itself.

33. CALCULATION.

THE MENTAL ARITHMETICIAN. — Numerical computation ability to reckon figures in the head; cognizance and memory of numbers; mental arithmetic.

Adapted to numerical relations.

SMALL.

LARGE.

No. 179. — MATHEMATICIAN.[1]   No. 180.

VERY LARGE. — Possess this calculating capability in a most extraordinary degree; can add several columns at once very rapidly and correctly, and multiply and divide with the same intuitive powers; love mental arithmetic exceedingly, and with large reflectives are a natural mathematician: p. 203.

[1] Zerah Colburn, at the age of nine years, without education, astonished the world by his great calculating talent. George Combe, though he studied mathematics seven years, never could master the multiplication table.

LARGE. — Excel in mental arithmetic, in adding, subtracting, multiplying, dividing, reckoning figures, casting accounts, etc., in the head; with large perceptives, have excellent business talents; and large Locality and Causality added, excel in mathematics : p. 202.

FULL. — Possess good calculating powers; with practice, can calculate in the head or by arithmetical rules easily and accurately, yet without practice are not remarkable; with large Form, Size, Comparison, Causality, and Constructiveness, can be a good geometrician or mathematician, yet will do better in the higher branches than merely the arithmetical : p. 204.

AVERAGE. — Can learn arithmetic and do quite well by practice, yet are not naturally gifted in reckoning figures : p. 202.

MODERATE. — Add, subtract, divide, and calculate with difficulty; and with large Acquisitiveness and perceptives, will make a better salesman than book-keeper : p. 204.

SMALL. — Are dull and incorrect in adding, subtracting, dividing, etc.; dislike figuring; are poor in arithmetic, both practical and theoretical, and should cultivate this faculty : p. 205.

VERY SMALL. — Can hardly count, much less calculate : p. 205.

To CULTIVATE. — Add, subtract, divide, multiply, count, and reckon figures in the head as far as possible, and learn and practice arithmetic.

To RESTRAIN, rarely ever necessary ; avoid counting things.

34. LOCALITY.

THE TRAVELLER. — Cognizance and recollection of places, roads, scenery, position, etc. ; desire to see places, ability to find them ; the geographical faculty ; keeping points of compass.

Adapted to nature's arrangement of space and place. Perverted, it creates a cosmopolitan disposition, and would spend everything in travelling.

VERY LARGE. — Always keep a correct idea of positions relative and absolute in the deep forests and winding street ; cannot be lost; are perfectly enamored of travelling ; have a passion for it : p. 206.

LARGE. — Remember the whereabouts of whatever is seen ; can carry points of the compass easily in the head, and are lost with difficulty either in the city, woods, or country ; desire to see places, and never forget them ; study geography and astronomy with ease ; rarely forget where things are seen; with Constructiveness, remember the arrangement of the various parts of a machine; with Individuality, Eventuality, and Human Nature, love to see men and things as well as places, and hence have a passion for travelling: p. 205.

FULL. — Remember places well, yet not extraordinarily so; can generally find the way, yet may sometimes be lost or confused; with large Eventuality, remember facts better than places: p. 207.

AVERAGE. — Recollect places and positions seen several times, yet in city or roads are occasionally lost; have no great geographical talent, yet by study and practice can do tolerably well : p. 205.

MODERATE. — Recollect places rather poorly; dare not trust to local memory in strange places or large cities; are not naturally good in geography, and to excel in it must study hard; should energetically cultivate this faculty by localizing everything, and remembering just how things are placed : p. 207.

SMALL. — Are decidedly deficient in finding places, and recollect them with difficulty even when perfectly familiar with them: p. 208.

VERY SMALL. — Must stay at home unless accompanied by others, because unable to find the way back : p. 208.

TO CULTIVATE. — Notice, as you go, turns in the road, landmarks, and objects by the way, geography and the points of compass, when you see things; and charge your memory where on a page certain ideas or accounts stand recorded, and position in general; and study geography by maps and travelling, the location of anatomical and phrenological organs, and position or place in general : 467.

TO RESTRAIN. — Settle down, and give up roving and travelling.

LITERARY FACULTIES.

These recollect information and anecdotes, and remember matters of fact and knowledge in general, and give what is called a good memory.

Adapted to facts, dates, and communicating ideas and feelings.

VERY LARGE. — Have a most remarkable memory; are extraordinarily well-informed, if not learned and brilliant; according to advantages are a first-rate scholar; have a literal passion for literary pursuits, and are remarkably smart and intelligent.

LARGE. — Are knowing, and off-hand; can show off to good ad

vantage in society; with large Ideality, are brilliant as well as talented; have an excellent memory.

FULL. — Have a fair matter-of-fact cast of mind and knowing powers, fair scholarship, and a good general memory.

AVERAGE. — If cultivated, have a good memory, and store up considerable knowledge; yet, without cultivation, only a commonplace memory and no great knowledge.

MODERATE. — Know more than you can think of at the time, or tell; with large reflective faculties, have more judgment than memory, and strength of mind than ability to show off.

SMALL, OR VERY SMALL. — Have a poor memory of most things, and inferior literary capabilities.

TO CULTIVATE. — Read, study, inform yourself, read the papers; keep pace with the improvements of the day; study history and the experimental sciences; and pick up and store up whatever kinds of knowledge, in your line of business, and of matter-of-fact knowledge, comes in your way; write your thoughts in a daily journal, or for the press; join a lyceum or debating society, and read history and science with a view to remember all you read and know, for the purpose of using it in argument; remember the news, and tell it to friends; in short, read, write, and talk.

TO RESTRAIN. — Read and study less; divert your mind from books and business by cultivating the other, and especially physical faculties, and never read, or study, or write nights.

35. EVENTUALITY.

THE HISTORIAN. — Memory of facts; recollection of circumstances, news, occurrences, and historical, scientific, and passing events — what has been said, seen, heard, and once known.

Adapted to action, and those changes constantly occurring around and within us.

VERY LARGE. — Are smart, bright, and knowing in the extreme; possess a wonderfully retentive memory of everything like facts and incidents; with large Language and Imitation, tell a story admirably, and excel in fiction, etc.: have a craving thirst for knowledge, and iterally devour books and newspapers, and never forget anything once seen or known: p. 211.

LARGE. — Have a clear and retentive memory of historical facts,

 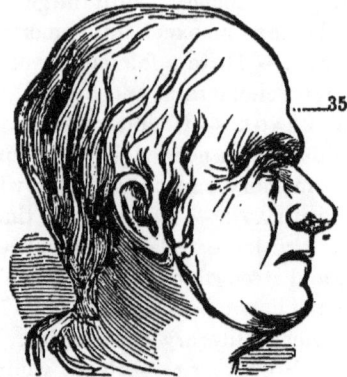

No. 181. — LARGE.   No. 182. — SMALL.

general knowledge, what has been seen, heard, read, done, etc., even in detail; considering advantages, are well informed and knowing; desire to witness and institute experiments; find out what is and has been, and learn anecdotes, particulars, and items of information, and readily recall to mind what has once entered it; have a good general matter-of-fact memory, and pick up facts readily; with Calculation and Acquisitiveness large, remember business matters, bargains, etc.; with large social feelings, recall friends to mind, and what they have said and done; and with large Locality, associate facts with the place where they transpired, and are particularly fond of reading, lectures. general news, etc., and can become a good scholar: p. 210.

FULL. — Have a good general memory of matters and things, yet it is considerably affected by cultivation — that is, have a good memory if it is habitually exercised, but if not, only an indifferent one; with large Locality, recollect facts by associating them with places, or where on a page they are narrated; with large reflectives, remember thoughts better than facts, and facts by associating them with their principles; and with large Language, tell a story quite well: p. 212.

AVERAGE. — Remember leading events and interesting particulars, yet are rather deficient in memory of items and details, except when it is well cultivated: p. 209.

MODERATE. — Are rather forgetful, especially in details; and with moderate Individuality and Language, tell a story very poorly, and should cultivate memory by its exercise: p. 212.

SMALL. — Have a treacherous and confused memory of circumstances; often forget what is wanted, intended to be said, done, etc.; have a poor command of knowledge, are unable to swear positively to details, and should strenuously exercise this memory: p. 213.

VERY SMALL. — Forget almost everything: p. 213.

To CULTIVATE. — *Charge* your mind with whatever transpires; remember what you read, see, hear, and often recall and reimpress it, so that you could swear definitely in court; impress on your mind what you intend to do and say at given times; read history, mythology, etc., with a view to weave such knowledge into every-day life; tell anecdotes; recount incidents in your own life, putting in all the little particulars; write down what you would remember, yet only to impress it, but trust to memory, not to manuscript: 476.

To RESTRAIN. — Read less; never allow yourself to recount the painful vicissitudes of life, or to renew past pain by remembrance, for this only does damage; but when you find your mind running on painful subjects, change it to something else, and try to forget whatever in the past is saddening.

36. TIME.

THE INNATE TIME-KEEPER. — Periodicity; cognizance and recollection of duration, succession, the lapse of time, when things occurred, etc.; ability to carry the time of the day in the head, tell when, how long, etc.; punctuality.

Adapted to Nature's times and seasons. Perverted, it is excessively pained by not keeping time in music, steps, etc.

VERY LARGE. — Can wake up at any preappointed hour, tell the time of day by intuition almost as correctly as with a time-piece, and the time between events, and are a natural chronologist: p. 216.

LARGE. — Can generally tell when things occurred, at least the order of events, and the length of time between one occurrence and another, etc.; tell the time of day well, without time-piece or sun; and keep an accurate mental chronology of dates, general and particular; with large Eventuality, rarely forget appointments, meetings, etc., and are a good historian, and always punctual: p. 215.

FULL. — With cultivation, can keep time in music, and also the time of day in the head quite correctly, yet not remarkably: p. 216.

AVERAGE. — With practice, have a good memory of dates and successions, yet without it are rather deficient: p. 214.

MODERATE.— Have a somewhat imperfect idea of time and dates and with moderate Eventuality and Language, are a poor historian p. 216.

SMALL.— Fail to keep the correct time in the head, or awaken at appointed times; have a confused and indistinct idea of the time when things transpired, forget dates, and lack punctuality : p. 217.

VERY SMALL.— Are almost destitute of this faculty : p. 217.

To CULTIVATE.— Periodize everything; rise, retire, prosecute your business, everything, by the clock ; appropriate particular times to particular things, and deviate as seldom as possible; in short, cultivate perfect regularity in all your habits, as respects time · 491.

To RESTRAIN.— Break in upon your tread-mill monotony, and deviate now and then, if only for diversion, from your routine.

37. TUNE.

THE INTUITIVE MUSICIAN.— Musical instinct, inspiration and genius ; ability to learn and remember tunes by rote.

Adapted to the musical octave.   Perversion — excessive fondness for music to the neglect of other things.

VERY LARGE.— Possess extraordinary musical taste and talent, and are literally transported by good music ; and with large Imitation and Constructiveness, fair time, and a fine temperament, are an exquisite performer ; learn tunes by hearing them sung once; sing in spirit and with melting pathos; show intuitive taste and skill; sing *from* the soul and *to* the soul: p. 219.

LARGE.— Love music dearly; have a nice perception of concord, discord, melody, etc., and enjoy all kinds of music; with large Imitation, Constructiveness, and Time, can make most kinds, and play well on musical instruments; with large Ideality, impart a richness and exquisiteness to musical performances ; have a fine ear for music and are tormented by discord, but delighted by concord, and take a great amount of pleasure in the exercise of this faculty; with large Combativeness and Destructiveness, love martial music; with large Veneration, sacred music; with large Adhesiveness and Amativeness, social and parlor music; with large Hope, Veneration, and disordered nerves, plaintive, solemn music, etc. : p. 218.

FULL.— Have a good musical ear and talent; can learn tunes by rote quite well; and with large Ideality and Imitation, can become a good musician, yet will require practice : p. 220.

AVERAGE. — Have fair musical talents, yet, to be a good musician, require considerable practice; can learn tunes by rote, yet with some difficulty; with large Ideality and Imitation, may be a good singer or player, yet are indebted more to art than nature; show more taste than skill; and love music better than can make it : p. 217.

MODERATE. — Have moderate taste and talent for music, yet, aided by notes and practice, may sing and play quite well, but will be mechanical, and lack that pathos which reaches the soul : p. 220.

SMALL. — Learn to sing or play tunes with great difficulty, and that mechanically, without emotion or effect : p. 221.

VERY SMALL. — Have scarcely any musical idea or feeling, so little as hardly to tell Yankee Doodle from Old Hundred : p. 221.

To CULTIVATE. — Try to sing; learn tunes by ear; and practice vocal and instrumental music : 504.

To RESTRAIN. — Give relatively less time and feeling to music, and more to other things.

38. LANGUAGE.

THE TALKER. — The expression of all mental operations by words, written or spoken, by gestures, looks, and actions ; the communicating faculty and instinct.

Adapted to man's requisition for holding communication with man.. Perversion — verbosity, pleonasm, circumlocution, garrulity, excessive talkativeness, telling what does harm, etc.

VERY LARGE. — Are exceedingly expressive in all said and done ; have a most expressive countenance, eye, and manner in everything ; and emphatic way of saying and doing everything, and thoroughly impress the various operations of your own minds on the minds of others ; use *the very* word required by the occasion ; are intuitively grammatical, even without study, and say oratorically whatever you attempt to say at all ; commit to memory by reading or hearing once or twice ; learn languages with remarkable facility ; are both fluent and copious, even redundant and verbose ; with large or very large Imitation, add perfect action, natural language, and gesticulation to perfect verbal selection ; with large Ideality, are elegant and eloquent ; and with large Individuality, Eventuality, Comparison, and organic quality added, possess natural speaking talents of the highest order ; say the very thing, and in the very best way ; choose words

LARGE.

SMALL.

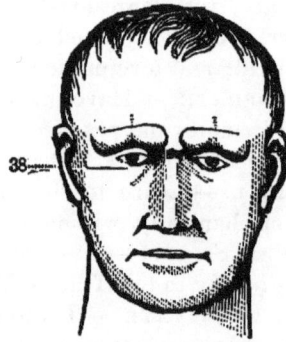

No. 183. — CHARLES DICKENS.

No. 184. — BRUNEL.

almost as by inspiration, and evince the highest order of communicating capacity : p. 226.

LARGE. — Express ideas and feelings well, both verbally and in
writing ; can learn to speak languages easily ; recollect words and
commit to memory well ; have freedom, copiousness, and power of expression ; with large Amativeness, use tender, winning, persuasive
words ; with large Combativeness and Destructiveness, severe and
cutting expressions ; with large moral faculties, words expressive of
moral sentiments ; with large Acquisitiveness, describe in glowing
colors what is for sale ; with large Ideality, employ richness and
beauty of expression, and love poetry and oratory exceedingly ; with
large Imitation, express thoughts and emotions by gesticulation ; with
activity great and Secretiveness small, show in the looks the thoughts
and feelings passing in the mind ; with large reflective faculties,
evince thought and depth in the countenance ; with large Comparison, use just the words which convey the meaning intended ; with
large Ideality, Individuality, Eventuality. Comparison, and the mental temperament, can make an excellent editor or newspaper writer ;
and with large Causality added, a philosophical writer, etc. : p. 224.

FULL. — Say well what is said at all, yet are not garrulous ; with
small Secretiveness, speak without qualification, and also distinctly
and pointedly ; express the manifestations of the larger faculties with
much force, yet not of the smaller ones ; with large Secretiveness and
Cautiousness, do not always speak to the purpose, and make ideas

fully understood, but use rather non-committal expressions; with large Comparison, Human Nature, Causality, Ideality, activity, organic quality, and power, have first-rate writing talents, and can speak well, yet-large Secretiveness impairs speaking and writing talents by rendering them wordy and non-committal : p. 227.

AVERAGE. — Have fair communicating talents, yet not extra; with activity great and Secretiveness small, speak right out, and to the purpose, yet are not eloquent, and use commonplace words and expressions; with large Individuality, Eventuality, and Comparison, and moderate Secretiveness, can make an excellent writer by practice ; use none too many words, but express yourself clearly and to the point ; with large Causality, have more thought than language; with moderate Individuality and Eventuality, find it difficult to say just what is desired, and are not fully and easily understood; with large Ideality, have more beauty and elegance than freedom : p. 222.

MODERATE. — Are not particularly expressive in words, actions, or countenance, nor ready in communicating ideas and sentiments; with large Ideality, Eventuality, Comparison, activity, and power, may succeed well as a writer, yet not as a speaker; talk fast, but use only common language ; with large Causality and moderate-Eventuality, have abundance of thoughts, but find it quite difficult to cast them into sentences, or bring in the right adjectives and phrases at the right time; are good in matter, yet poor in delivery; commit to memory with difficulty, and fail to make ideas and feelings fully understood, and to excite like organs in others; with large Eventuality, Locality, Form, and Comparison, may be fair as a linguist, and learn to read foreign languages, yet learn to speak them with difficulty, and are barren in expression, however rich in matter : p. 228.

SMALL. — Have poor lingual and communicative talents; hesitate for words; speak with extreme difficulty and very awkwardly, and should cultivate this faculty by talking and writing much : p. 228.

VERY SMALL. — Can hardly remember or use words at all.

To CULTIVATE. — Talk, write, speak as much, eloquently, and well as you can; often change clauses in order to improving sentences ; erase unnecessary and improper words, and choose the very words exactly expressive of the desired meaning; throw feeling and expression into all you say; give action and expressiveness to countenance ; study languages and the classics, but especially fluency in your mother tongue ; narrate incidents; tell what you have heard, seen, read, done; debate; if religious, lead in religious exercises — anything, everything to discipline and exercise this faculty: 515.

To RESTRAIN. — Talk less; never break in when others are talking; lop off redundancies, pleonasms, and embellishments, and use simple instead of bombastic expressions.

REFLECTIVE OR REASONING FACULTIES.

These give a philosophizing, penetrating, investigating, originating cast of mind; ascertain causes and abstract relations; contrive, invent, originate ideas, etc. Adapted to the first principles, or laws of things.

VERY LARGE. — Possess extraordinary depth of reason and strength of understanding; and with large perceptives, extraordinary talents, and manifest them to good advantage; with perceptives small, have great strength of mind, yet a poor mode of manifesting it; are not appreciated, and lack intellectual balance, and are more plausible than reliable, and too deep to be clear.

LARGE. — Possess the higher capabilities of intellect; reason clearly and strongly on whatever data is furnished by the other faculties; have soundness of understanding, depth of intellect, and that weight which carries conviction, and contributes largely to success in everything; with perceptives small, possess more power of mind than can be manifested, and fail to be appreciated and understood, because more theoretical than practical.

FULL. — Possess fair reflective powers, and reason well from the data furnished by the other faculties; and with activity great, have a fair flow of ideas and good general thoughts.

AVERAGE. — Reason fairly on subjects fully understood, yet are not remarkable for depth or clearness of idea; with cultivation, will manifest considerable reasoning power — without it, only ordinary.

MODERATE. — Are rather deficient in power of mind; but with large perceptives, evince less deficiency of reason than is possessed.

SMALL. — Have inferior reasoning capabilities.

VERY SMALL. — Are almost destitute of thought, idea, and sense.

To CULTIVATE. — Muse, meditate, ponder, reflect on, think, study and pry deep into the abstract principles and nature of things.

To RESTRAIN. — Theorize less, and give more time to facts.

39. CAUSALITY.

LARGE.

15

SMALL.

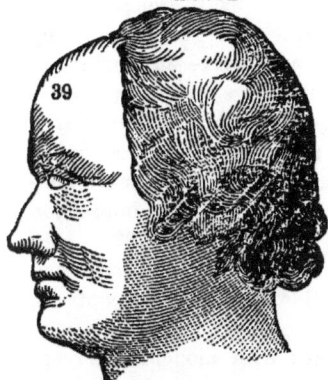

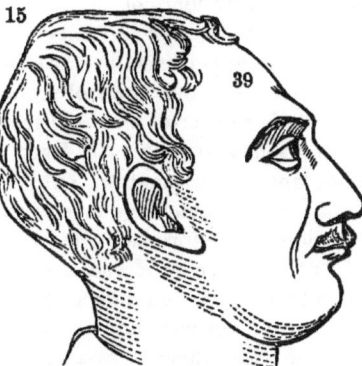

No. 185. — Dr. Gall.   No. 186. — Hewlett, Actor.

The Thinker and Planner. — Perception and application of causation ; reason ; deduction ; originality ; depth of thought ; forethought ; comprehensiveness of mind ; devising ways and means ; invention ; creating resources ; reasoning from causes to effects ; profundity.

Adapted to Nature's laws, plans, causes, and effects. Perverted, it reasons in favor of untruth and injurious ends.

Very Large. — Possess this cause seeking and applying power to an extraordinary degree ; perceive by intuition those deeper relations of things which escape common minds ; are profound in argument and philosophy, and deep and powerful in reasoning, and have great originality of mind and strength of understanding ; see Large : p. 236.

Large. — Desire to know the *whys* and *wherefores* of things, and to investigate their laws ; reason clearly and correctly from causes to effects ; have uncommon capabilities of planning, contriving, inventing, creating resources, and making head save hands ; kill two birds with one stone ; predicate results, and arrange things so as to succeed ; put things together well ; with large Combativeness, love to argue ; with large perceptives, are quick to perceive facts and conditions, and reason powerfully and correctly from them ; with Comparison and Conscientiousness large, reason forcibly on moral truths ; with the selfish faculties strong, will so adapt ways and means as to serve personal purposes; with moderate perceptives, are theoretical, and excel more in principles and philosophy than facts; remember laws better than details; with Comparison and Human Nature large, are particularly fond of mental philosophy, and excel therein; with Individuality and Eventuality only moderate, are guided more by reason than experience, by laws than facts, and arrive at conclusions more from reflection than observation; with large perceptives, possess a higher order of practical sense and sound judgment; with large Comparison and moderate Eventuality, remember thoughts, inferences, and subject-matter, but forget items; with the mental temperament and Language moderate, make a much greater impression by action than expressions, by deeds than words, etc. : p. 233.

FULL. — Have good cause seeking and applying talents; reason, and adapt ways and means to ends, well; with large perceptives, Comparison, activity, and organism, possess excellent reasoning powers, and show them to first-rate advantage; with moderate perceptives and large Secretiveness, can plan better than reason; with large Acquisitiveness and moderate Constructiveness, lay excellent money-making, but poor mechanical plans, etc. : p. 236.

AVERAGE. — Have only fair sense and judgment; plan and reason well in conjunction with the larger faculties, but poorly with the smaller; with moderate Acquisitiveness, lay poor money-making plans; but with large Conscientiousness, reason well on moral subjects, especially if Comparison is large, etc. : p. 231.

MODERATE. — Think little; rather lack discernment and causation; perceive causes when presented by other minds, yet do not originate them; with activity and perceptives large, may do well in ordinary business routine, yet fail in difficult matters : p. 237.

SMALL. — Are deficient in reasoning and planning power and sense; need perpetual telling and showing; seldom arrange things beforehand, and then poorly; should work under others; lack force of idea and strength of understanding : p. 238.

VERY SMALL. — Are idiotic in reasoning and planning : p. 238.

TO CULTIVATE. — First and mainly, study Nature's causes and effects, adaptations, laws, both in general and in those particular departments in which you may feel any special interest; think, muse, meditate, reason, cogitate; give yourself up to the influx of new ideas; plan; adapt ways and means to ends; endeavor to think up the best ways and means of overcoming difficulties and bringing about results; especially study Phrenology and its philosophy, for nothing ε equally suggestive of original ideas, or as explanative of Nature's laws and first principles : 545–548.

To RESTRAIN — which is rarely necessary — divert your mind from abstract thought by engaging more in the practical and real, nor allow any one thing, as inventing perpetual motion, or reasoning on any particular subject, to engross too much attention.

40. COMPARISON.

LARGE. SMALL.

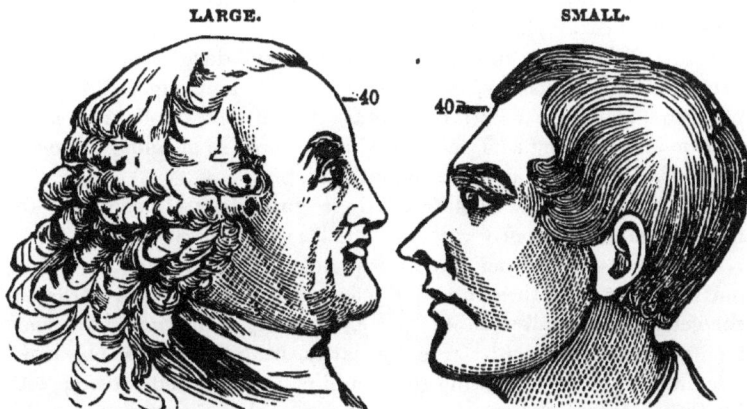

No. 187. — LINNÆUS. No. 188. — MR. BARLOW.

THE CRITIC. — Inductive reasoning; ability and desire to analyze, illustrate, classify, compare, and draw inferences.

Adapted to Nature's classifications of all her works. Perverted, is too redundant in proverbs, fables, and figures.

VERY LARGE. — Possess this analyzing, criticising, and inductive faculty in a truly wonderful degree; illustrate with great clearness and facility from the known to the unknown; explain things plausibly and correctly; discover the deeper analogies which pervade nature, and have an extraordinary power of discerning new truths; with large Individuality, Eventuality, and activity, have a great faculty of making discoveries; with large Language, use words in their exact meaning, and are a natural philologist : p. 243.

LARGE. — Reason clearly and correctly from conclusions and scientific facts up to the laws which govern them; discern the known from the unknown; detect error by its incongruity with facts; have an excellent talent for comparing, explaining, expounding, criticising, exposing, etc.; employ similes and metaphors well; put this and that together, and draw correct inferences from them; with large Continuity, use well-sustained figures of speech, but with small Continuity, drop the figure before it is finished; with large Individuality, Eventuality, activity, and power, have a scientific cast of mind; with large Veneration, reason about God and His works; with large Language, use words in their exact signification; with large Mirthfulness, strike the nail upon the head in all criticisms, and hit off the oddities of people to admiration; with large Ideality, evince beauty, taste, and propriety of expression, etc.: p. 241.

FULL. — Possess a full share of clearness and demonstrative power, yet with large Causality, and only moderate Language, cannot explain to advantage; with large Eventuality, reason wholly from facts; with moderate Language, fail in giving the precise meaning to words; and make fair analytical discriminations: p. 243.

AVERAGE. — Show this talent in a good degree along with the larger organs, but poorly with the smaller: p. 239.

MODERATE. — Rather fail in explaining, and clearing up points, putting things together, drawing inferences, and often use words incorrectly; with Individuality and Eventuality moderate, show much mental weakness; with large Causality, have fair ideas, but make wretched work in expressing them, and cannot be understood; with Mirthfulness full or large, try to make jokes, but they are always ill-timed and inappropriate: p. 244.

SMALL. — Have a poor talent for drawing inferences; lack appropriateness in everything, and should cultivate this faculty: p. 244.

VERY SMALL. — Have little, and show less sense: p. 244.

TO CULTIVATE. — Put this and that together and draw inferences; spell out truths and results from slighter data; observe effects, with a view to deduce conclusions therefrom; study logic and metaphysics, theology and ethics included, and draw nice discriminations; explain and illustrate your ideas clearly and copiously, and exercise it in whatever form circumstances may require: 536.

TO RESTRAIN. — Keep back redundant illustrations and amplifications, and base important deductions on data amply sufficient.

41. HUMAN NATURE.

THE PHYSIOGNOMIST. — Perception of character; discernment of motives; intuitive reading of men by minor signs

Adapted to man's need of knowing his fellow-men. Per-verted, it produces suspiciousness.

VERY LARGE. — Form a correct judgment as to the character of all, and especially of the opposite sex, at first sight, as if by intuition ; may always trust first impressions ; are a natural physiognomist; and with Agreeableness large, know just when and how to take men, and hoodwink ; with Secretiveness added, but Conscientiousness moderate, are oily and palavering, and flatter victims ; serpent-like, salivate before swallowing ; with Comparison and organic qual-ity large, dearly love to study human nature, practically and theoret-ically, and therefore mental philosophy, Phrenology, etc.

LARGE. — Read men intuitively from their looks, conversation, manners, walk, and other kindred signs of character ; with Individu-ality and Comparison large, notice all the little things they do, and form a correct estimate from them, and should follow first impressions respecting persons ; with full Secretiveness and large Benevolence, know just how to take men, and possess much power over mind ; with Mirthfulness and Ideality large, see faults, and make much fun over them ; with Comparison large, have a talent for metaphysics, etc.

FULL. — Read character quite well from the face and external signs, yet are sometimes mistaken ; may generally follow first impres-sions safely ; love to study character ; with Ideality and Adhesive-ness large, appreciate the excellences of friends ; with Parental Love large, of children ; with Combativeness and Conscientiousness very large, all the faults of people ; and with only average Adhesiveness, form few friendships, because detecting so many blemishes in others.

AVERAGE. — Have fair talents for reading men, yet not extra.

MODERATE. — Fail somewhat in discerning character ; occasion-ally form wrong conclusions concerning people ; should be more sus-picious, watch people closely, especially those minor signs of charac-ter dropped when off their guard ; make ill-timed remarks ; address people poorly ; often say and do things which have a different effect from that intended, etc.

SMALL. — Are easily imposed on ; think everybody tells the truth ; are too confiding, and fail in knowing where and how to take men.

VERY SMALL. — Know almost nothing about human nature.

TO CULTIVATE. — Scan closely all the actions of men, in order to ascertain their motives and mainsprings of action ; look with a sharp eye at man, woman, child, all you meet, as if you would read them

11 .

through; note particularly the expressions of the eye, as if you would imbibe what it signifies; say to yourself, What faculty prompted this expression and that action? drink in the general looks, attitude, natural language, and manifestation of men, and yield yourself to the impressions naturally made on you, that is, study human nature both as a philosophy and a sentiment, or as if being impressed thereby; especially study Phrenology, for no study of human nature at all compares with it, and be more suspicious: 540.

To RESTRAIN. — Be less suspicious, and more confidential.

42. AGREEABLENESS.

THE COURTIER. — Blandness; persuasiveness; pleasantness; complaisance; suavity; palaver; that which compliments.
Adapted to please and win others.

VERY LARGE. — Are peculiarly winning and fascinating in manners and conversation, and delight even opponents.

LARGE. — Have a pleasing, persuasive, and conciliatory address; with Adhesiveness and Benevolence large, are generally liked; with Comparison and Human Nature large, say unacceptable things in an acceptable manner, and sugar over expressions and actions.

FULL. — Are pleasing and persuasive in manner, and with Ideality large, polite and agreeable, except when the repelling faculties are strongly excited; with small Secretiveness, and strong Combativeness and activity, are generally pleasant, but when angry are sharp and blunt; with large Benevolence and Mirth, are good company.

AVERAGE. — Are fairly pleasant in conversation and appearance, except when the selfish faculties are excited, but are then repulsive.

MODERATE. — Rather lack the pleasant and persuasive, and should by all means cultivate them by smoothing over all said and done.

SMALL. — Say even pleasant things very unpleasantly, and fail sadly in winning the good graces of people.

To CULTIVATE. — Kiss the blarney stone; take lessons from "Sam Slick;" try to *feel* agreeably, and express those feelings in as pleasant and bland a manner as possible; study and practice politeness as both an art and a science; compliment what in others you can find worthy, and render yourself just as acceptable as you can: 300

RULES FOR FINDING THE ORGANS.

PHRENOLOGY is a science of FACTS. Observation discovered, and alone can perfect it; and is the grand instrumentality of its propagation. To be convinced of its truth, men require to see it proved by INDUCTION and experiment. Hence the importance of definite RULES for finding its organs, by which all can test its truth, and prosecute its study.

The best mode of investigating its truth is somewhat as follows: You know a neighbor who has extreme Firmness in character, and is as obstinate as a mule. Now, learn the location of the phrenological organ of Firmness (see cut No. 159, at 16), and see whether he has this organ as conspicuous as you know this faculty is in his character, and if so, you have a strong phrenological fact.

You know another neighbor who is exceedingly cautious, timid, safe, wise, and hesitating; who always looks at objections and difficulties, instead of at advantages; takes abundant time to consider, and procrastinates: now learn the location of Cautiousness (see cut No. 155 at 13), and see whether he has this phrenological organ as conspicuous as you know he possesses this faculty in his character. By pursuing this course, you can soon obtain a sure knowledge of the truth or falsity of phrenological science. This is also altogether the best mode of convincing unbelievers of its truth. The intelligent cannot resist proof like this.

To promote the application of this science, we give the following RULES FOR FINDING ITS ORGANS. Follow these rules exactly, and you will have little difficulty in locating at least all the prominent ones, and from them decipher that of the others.

THE TEMPERAMENT should be noticed first, that is, the *organization* and physiology, with this principle for your basis: that when the bodily texture or form is coarse, or strong, or fine, or soft, or weak, or sprightly, the texture of the brain will correspond with that of body, and the mental characteristics with that of the brain.

The ruling faculties should be observed next. In phrenological language see what faculties PREDOMINATE. The relative *size* of organs does not always determine this point. Some faculties, though predominant, cannot, in their very nature, constitute a motive for action, but are simply executive, carrying into effect the dominant motives. For example, Combativeness rarely ever becomes a distinct motive for action. Few men love simply to struggle, quarrel, or fight for fun, but exercise Combativeness merely as a means of obtaining the things desired by the other dominant faculties. Few men live merely to exercise will; that is, Firmness generally carries into effect the desires of the other faculties, and simply keeps them at their work, and thus of some other faculties; whereas, Amativeness, Friendship, Alimentiveness, Acquisitiveness, Benevolence, Veneration, Conscientiousness, Intellect, Constructiveness, Ideality, and the observing faculties become dominant motives. And it requires much phrenological shrewdness to ascertain what faculty, or combination of faculties thus controls the character.

Starting at the outer angle of the eye, draw a line to the middle of the top of the ears, and Destructiveness (see cut No. 148) is exactly under this point, and extends upward about half an inch above the top of the ears. In proportion to its size will the head be wide between the ears. When Secretiveness is small and Destructiveness large, there will be a horizontal ridge extending forward and backward, more or less prominent, according to the size of this organ and a hollow right above.

Secretiveness is located three quarters of an inch above the middle of the top of the ears. When this organ is large, it rarely gives a distinct projection, but simply rounds out the head at this point (cut No. 148 at 9). When the head widens rapidly from the junction of the ears as you rise upward, Secretiveness is larger than Destructiveness; but when the head becomes narrower as you rise, it is smaller than Destructiveness. It is small in cut No. 155.

To find these two organs, and their relative size, place the third finger of each hand upon the head, just at the top of the ears; let the lower side of the third finger be even with the upper part of the ear; that finger then rests upon Destructiveness. Then spread the second finger about an eighth of an inch from the other, and it will rest upon Secretiveness. Let the end of your longest finger come as far forward as the fore part of the ears, and they will then rest upon these two organs.

Extend this same line straight backward an inch and a half o three quarters, and you are on Combativeness (cut No. 147 at 8) This organ starts at the middle of the back part of the ears, and runs upward and backward toward the crown of the head.   To ascertain its relative size, steady the head with one hand, say the left, and place the balls of your right fingers upon the point just specified, letting your elbow be somewhat below the subject's head, which will bring your fingers directly ACROSS the organ.   Its size may be ascertained partly from the general fullness of the head, and partly from its sharpness, according as the organ is more or less active ; yet observers sometimes mistake this organ for the mastoid process directly behind the lower part of the ears.   Remember our rule, namely, a line drawn from the outer angle of the eye to the top of the ear, and continued an inch and a half straight back.

To find PARENTAL LOVE (cut No. 143 at 3), extend this line straight back to the middle of the back head, and, in proportion as the head projects backward behind the ears at this point, will this organ be larger or smaller.

About an inch BELOW this point is the organ which controls MUSCULAR MOTION ; and in proportion as this occipital process is more or less prominent, will the muscular system be more or less active and powerful.   Those who have it large will be restless, always moving a hand or foot when sitting, and even when sleeping ; will be light-footed, easy-motioned, fond of action, and willing to work, as well as possessed of a first-rate constitution.   But those who have no prominence will be found proportionally inert.

INHABITIVENESS is located three fourths of an inch ABOVE Parental Love (cut No. 145 at 3).   When it is large, and Continuity moderate, there will be found a prominence somewhat resembling the fore part of a flat-iron, at the middle of the head, together with a sharp prominence at this point ; but when Inhabitiveness is small, there will be a depression just about large enough to receive the end of a finger, with the bow downward.

FRIENDSHIP is an inch on each side of this point.   When it is large, especially if Inhabitiveness and Continuity are small, there will be two swells, somewhat resembling the larger end of an egg ; but if small, the head will retire at this point.

CONTINUITY (cut No. 146 at 6) is located directly above Inhabitiveness and Friendship.   Its deficiency causes a depression resembling a new moon, with the horns turning DOWNWARD, surrounding the organs of Inhabitiveness and Friendship. When Continuity is large. however. there will be no swell, but only a FILLING OUT of the head at this point.

AMATIVE·ESS is large in cut No. 141 at 1, but small in No. 143. Take the middle of the back part of the ears as your starting-point ; draw a line backward an inch and a half, and you are upon this organ. Yet the outer portion next to the ear is more animal, while its inner portion is more platonic.

To find CAUTIOUSNESS (cut No. 155), draw a perpendicular line, when the head is erect, from the extreme back part of the ear, straight up the sides of the head, and just where it begins to round off, to form the top. Cautiousness is located there, is generally well developed in the American head, and those prominences generally seen at this point are caused by a full development of this organ.

ALIMENTIVENESS (cut No. 150 at 10) is half an inch forward, inclining a little downward, of the upper junction of the ear with the head. Then rise three quarters of an inch straight upward, and you are on that part of ACQUISITIVENESS which *gets* property. Or thus · Take the middle of the top of the ear as your starting-point ; draw a perpendicular line an inch upward, and you are on Secretiveness ; then about an inch forward is Acquisitiveness. It is very large in cut No. 152, but very small in No. 153 at 11. When the head widens rapidly as you pass from the outer angles of the eyes to the top of the ears, Acquisitiveness is large ; but when the head is thin in this region, Acquisitiveness is small.

SUBLIMITY, IDEALITY (cut No. 133). and CONSTRUCTIVENESS (22 in No 168), can be found by first finding Cautiousness ; then pass directly forward an inch, and you are on Sublimity ; extend this line another inch, and you are on Ideality ; then an inch downward brings you upon Constructiveness.

It should be remembered that Cautiousness, Sublimity, and Ideality are just upon the round of the head, or between its top and sides. Usually the head is much wider at Cautiousness than at Sublimity, and at Sublimity than Ideality. When, however, the head is as wide at Ideality as at Cautiousness, the subject will possess unusual good taste, purity, refinement, elevation, and personal perfection. Half an inch forward of Ideality is the organ which appertains to dress, and gives personal neatness. In those who care but little what they wear or how they appear, this organ will be found small.

FIRMNESS (at 16 in cut No. 159) can best be found by letting the subject be erect, and taking the opening of the ear as your starting-point; draw a line straight upward till you reach the middle line on the top of the head, and you are on the fore part of Firmness. When this organ is large, and Veneration small, its forward termination resembles in shape the fore part of a smoothing-iron, rapidly widening as it runs backward. The organ is usually about an inch and a half long. It is very large in cut No. 164, but moderate in No. 165.

SELF-ESTEEM-(cut No. 140) is an inch and a half back of Firmness. Its upper part gives a lofty, aspiring air, magnanimity, and a determination to do something worthy; while half an inch farther back is that part of Self-esteem which gives WILL, love of liberty, and a determination not to be ruled. It is large in cuts Nos. 140, 145, and 159.

APPROBATIVENESS (cut No. 158) is located on the two sides of Self-esteem, about an inch outwardly. These two lobes run backward toward Friendship, and upward toward Conscientiousness.

The relative size of Approbativeness and Self-esteem may be found thus : Placing the left hand upon the forehead, find Firmness ; and then move it two inches directly backward, and place the balls of the second and third fingers on this point. When Self-esteem is small, these balls will fall into the hollow which indicates its deficiency, while the ends of the fingers will strike upon the swells caused by Approbativeness, when this organ is large; and the middle of the second joint of these fingers will admeasure the size of that lobe of Approbativeness which is next to it. Or thus : Stand behind the patient, and so place your fingers upon his head that the second finger shall reach upward to the back part of Firmness ; then lay the first and second joints of that finger evenly with the head, and place the first and third fingers upon the head alongside of it. When Self-esteem is larger than Approbativeness, the second finger will be higher than the others ; but when its two lobes are larger than Self-esteem, the second finger will fall into a hollow running up and down, while the first and third fingers will rest upon them. Or thus : In nineteen females out of every twenty, Approbativeness will be found considerably larger than Self-esteem; and by applying this rule to their heads, a hollow will generally be found at Self-esteem, and a swell at Approbativeness, by which you can localize these organs ; and a few applications will soon enable you to form correct ideas of their appearance when large and small.

HOPE and CONSCIENTIOUSNESS (see cuts Nos. 162, 163, 16, 17) are found thus : That line already drawn to find Firmness passes over the back part of Hope, which is on each side of the fore part of Firmness, while Conscientiousness is just back of that line, on the two sides of the back part of Firmness, and joins Approbativeness posteriorly

As these two organs run lengthwise from Firmness down toward Cautiousness, and are near together. it is sometimes difficult to determine which is large and which small. The upper part of Conscientiousness, next to Firmness. experiences feelings of obligation to God, or sense of duty to obey his laws; while the lower part creates a feeling of obligation to our fellow-men.

VENERATION (large at 20 in cut No 164, but small in No. 165) is on the middle of the top head, or about an inch forward of Firmness; while BENEVOLENCE (large at 21 in cut No. 166, but small in No. 167) is about an inch forward of Veneration. When the middle of the top head rounds out and rises above Firmness and Benevolence, as in 164, Veneration is *larger* than either of these organs ; but when there is a swell at Benevolence. and a depression as you pass backward in the middle of the head, which rises again as you pass still further back to Firmness, Veneration is *smaller* than Benevolence or Firmness. The back part of Benevolence experiences philanthropy and a desire to do good and remove suffering on a large scale, while the fore part sympathizes, and bestows minor gifts in the family and neighborhood. The fore part of Veneration gives respect for our fellow-men, while the back part supplicates and depends upon the Deity. The fore part of Firmness, working with Conscientiousness, gives moral decision; while the latter, acting with Self-esteem, gives physical decision, determination to accomplish material objects, perseverance, etc., etc.

SPIRITUALITY is located on each side of Veneration. It may be found by standing behind the seated subject ; so place your fingers that the first fingers of each hand shall be about an inch apart — that the ends of your second fingers shall be about three quarters of an inch forward of a line drawn across the middle of the head from side to side, and the balls of your fingers will be on Spirituality. Or, reversing your position, so as to stand in FRONT of the subject, so place your hands that the first fingers of each hand shall be as before. about an inch apart, and the ends of your longest fingers shall just touch the fore part of Hope, and the balls of your second and third fingers will rest on Spirituality. This organ is generally low, so that it may usually be found by that *depression* which indicates its smallness. When it is large, the head is filled out in this region, instead of sloping rapidly from Veneration. Its two lobes are about an inch on each side of Veneration, and directly above Ideality.

IMITATION is large in cuts Nos. 171, 173, upon the two sides of Benevolence, but small in 172 and 174, directly forward of Spirituality. To find it, stand in front of the subject, place your hands so that the index fingers of each hand shall be separated about three quarters of an inch, and the end of your longest finger shall reach a line drawn across the middle of the head and the balls of your fingers will be on Imitation. It is found larger in children than adults; so that the ridge usually found in their heads at this point may be taken as the location of this organ. It runs from Benevolence downward toward Constructiveness. The upper part, toward Benevolence, mimics; the lower part, toward Constructiveness, makes after a pattern, copies, etc.

In the intellectual lobe, take the root of the nose as your starting-point; the first organ met in passing upward is INDIVIDUALITY (27 in cut No. 175). It is between the eyebrows, and when large causes them to arch DOWNWARD at their inner termination, and that part of the head to project forward. It is small in No. 176.

EVENTUALITY is very large in cut No. 181, but small in 182, just below the centre of the forehead, and in children is usually large, but frequently small in adults. From this centre of the forehead, COMPARISON (40 in cut No. 187) extends upward to where the head begins to slope backward to form its top; at which point, or between Benevolence and Comparison, HUMAN NATURE is located, which is usually large in the American head, as is also Comparison. AGREE-ABLENESS is located about an inch on each side of the organ of Human Nature, and is usually small, so that we can ascertain its location by observing its deficiency. When both of these organs are large, the forehead will be wide and full as it rounds backward to form the top head, or where the hair makes its appearance, as in cut No. 177, but it is small in 174. CAUSALITY, in cuts Nos. 173, 174, is located about an inch on each side of Comparison; and MIRTHFULNESS (large in cut No. 177, small in 178) about three quarters of an inch still further outwardly, toward Ideality. FORM, in cut No. 178, is located internally from Individuality, just above and partly between the eyes, so as to set them wider apart, in proportion as it is the larger.

SIZE is located just in the turn between the nose and eyebrows, or beneath the inner portion of the eyebrows; and when large, causes their inner portions to project forward over the inner portion of the eyes like the eaves of a house, giving to the eyes a sunken appearance. Its size can generally be observed by sight, yet if you would aid sight by touch, place the end of your thumb against the bridge of the nose, with your thumb nearly parallel with the eyebrows, and its ball will be upon Size. When this organ is large, it resembles half a bean.

To find WEIGHT and COLOR, let the eyes look straight forward, and draw an imaginary line from the middle of the eye to the eyebrow: Weight is located *internally* beneath the eyebrows, while Color is located *outwardly* from this line. ORDER is located just externally to Color, and TIME partly above and between Color and Order.

CALCULATION (33 in cut No. 179) is located beneath the outer termination of the eyebrows, and in proportion as they are long and extend backward of the eye, will this organ be more or less developed. It is small in cut No. 180. Three fourths of an inch ABOVE the outer angle of the eyebrow TUNE is located. Spurzheim's rule for finding it is this: Stand directly before the subject, and if the head widens over the outer end of the eyebrow as you rise upward, Tune is large, but a hollow at this point indicates small Tune. It is difficult to find its relative size in adults, but in children easy. Time and Tune join each other, and with Mirthfulness occupy the three angles of a triangle, nearly equilateral, the shortest side being between Time and Tune.

LANGUAGE is large in cut No. 183, but small in 184, and located partly above and behind the eyes. When large, it pushes the eyes downward and outward, and of course forward, which gives them a full and swollen appearance, as if they were standing partly out of their sockets, causing both the upper and under eyelids to be wide and broad. When the eyes are sunken, and their lids narrow, Language is small.

By following these rules exactly and specifically, the precise location of the organs can be ascertained, and a few observations upon heads will soon teach you the appearance of the respective organs when they are large, small, or medium. Some slight allowances must be made in calculating the size of the head, or the absolute size of the organs. Thus, the larger Combativeness is, the longer the line from Combativeness to the ear; yet large and small Combativeness do not vary this line over from a quarter to half an inch.

Probably the most difficult point of discrimination is between Hope and Conscientiousness. Hope is generally placed too far forward. Between Hope, Cautiousness, and Approbativeness there probably exists an organ, the natural functions of which are discretion. It measures words and acts, and in business leads one to take receipts, draw writings, etc. There are doubtless other organs yet undiscovered, especially in the middle line of the head, between Benevolence and Parental Love, and also between Imitation and Causality. Phrenology is yet in its infancy. Though it is perfect in itself, yet our KNOWLEDGE of it is not yet perfected. As every successive generation makes advances upon the preceding in astronomy, chemistry, and other departments of science, so Gall and Spurzheim have discovered only the landmarks of this science, leaving much to be filled up by us and those who come after us.

THE MATRIMONIAL ADAPTATION.

THE LOVE TASTES of men and women differ even more than their other tastes. " What is one's meat is another's poison." One man likes, another dislikes, the same qualities in the same woman, and thus of women. This natural law governs these tastes: those in either *extreme* in any respect love those best who are in an *opposite* extreme, while those who are *medium* in any quality affiliate best with those who are near themselves. Thus very large men love small women, and small men large women, while average men like average women best, yet can affiliate with either large, medium, or small; and so of women. Bright red hair prefers jet black, while medium can love medium, or black, or red ; and thus of curls. Tall persons should marry short, and slim, stocky ; while those medium in height may marry either or medium. Those having prominent noses and retiring chins and foreheads should select straight profiles, square faces, and high and wide foreheads, and large noses medium or small, Roman, pug, etc.

The impulsive love the calm, yet those who are neither may select either; and this principle applies to all the phrenological faculties. See this whole subject of adaptations and the sexual relations thoroughly discussed in the author's work on " Sexual Science and Restoration, or Manhood, Womanhood, Love, Selection, Courtship. Married Life, Reproduction, Paternity, Maternity, Puberty, Sexual Ailments, Beauty, etc. To know your matrimonial adaptation is so important that we append to our table a column for its record. Suit yourself as to those conditions unmarked, but select one near those marked.

CONTENTS.

———◆———

PROF. O. S. FOWLER'S REVISED WORKS.

COMPLETE IN TWO VOLUMES.

"Human Science; or, Phrenology; its Principles, Proofs, Faculties, Organs, Temperaments, Combinations, Conditions, Teachings, Philosophies, &c.; applied to Health; its Value, Laws, Functions, Organs, Means, Preservation, Restoration, &c.; Mental Philosophy, Human and Self-Improvement, Civilization, Home, Country, Commerce, Rights, Duties, Ethics, &c.; God, His Existence, Attributes, Laws, Works, Worship, Natural Theology, &c.; Immortality, its Evidences, Conditions, Relations to Time, Rewards, Punishment, Sin, Faith, Prayer, &c.; Intellect, Memory, Juvenile and Self-Education, Literature, Mental Discipline, the Senses, Sciences, Arts, Avocations, a Perfect Life, &c., &c." *

Man, Know Thyself, is the motto for the *race! Anthropology* is universal philosophy; because man is the epitome of the universe; while Mentality is the ultimate of man: therefore mental philosophy is the summary of all science. of all utility. Its study shows *how to live;* resolves all the problems of humanity; reveals those mental *fountains* from which all feelings and actions emanate; discloses the perfect man, and thereby shows all communities, all individuals, just wherein and how far each conforms thereto, and departs therefrom; and must therefore soon become *the great study of the race,* and so remain, "till time shall be no longer."

This Standard Work on Phrenology, by unfolding its principles, classifying its facts, giving its history and recent discoveries and improvements, embodying the gist of all its previous writings, and being a repository of whatever is known concerning it, becomes an unequalled public and personal benefaction none can afford to ignore. Reader, your *soul,* that sentient entity which alone enjoys, suffers, lives forever, constitutes existence, and is Jehovah's crowning work, deserves your supreme attention. Should and would you not analyze its component parts, and learn how their united action creates all your ever-varying functions, capacities, and virtues? Is examining a complicated machine anything, and is studying you own wonder-working *mind* nothing? What knowledge is a tithe as valuable as *self-knowledge?* What other can be turned to a hundredth part as good practical account? Is Nature a blank to you, and *human* nature a barren waste? Your inner *self-hood* is unfolded only in this work.

Its original theories of organic formation, showing *how* all structures become adapted to the specific requirements of each; of pain and punishment as necessarily remedial; of that motive power which propels the blood, heavenly bodies, &c.; of domestic architecture, the octal vs. decimal arithmetical system, &c., merit attention. In short, it combines a complete exposition of *all* the *departments* of man's body and mind, in one collective whole, and, with "Sexual Science," embodies all the works, writings, reflections, observations, and professional experiences of *half a century.* on *four generations,* of its Author, revised, enlarged, systematized, and condensed into one comprehensive work, amply illustrated by *over two hundred eye-teaching engravings,* which enable amateurs to commence and prosecute this study *without further aid;* yet it also elaborates its philosophies. and applies all to self and juvenile culture, and a perfect life, — ends how infinitely exalted! It naturally subdivides itself into six parts.

* All my chart-marked patrons, by transferring their markings to its table, can read their characters, in full, and learn from it how to cultivate themselves and children.

PART I. The Organism, discusses man's organic relations generally, including the fundamental principles of life: the construction of the mind and brain; the principles, proofs, facts, and history of Phrenology; the Temperaments, &c., as seen in its following

Synopsis. Value of life. Amount of happiness possible to all. Its improvement. Enjoying life *as we go*. Mind makes the man. Self-acting, natural laws, which are divine commands, cause all happiness, all suffering, and should be studied. All pain curative. All functions manifested by organs. All organic states similarly affect their functions. Body, brain, and mind in reciprocal sympathy. Bodily impairments create sinful proclivities. Materialism. Normal, abnormal, and harmonious action.

Phrenology and its Faculties defined. The brain, and its structure. The organ of the mind and body. Composed of separate organs. Size indicates power. Human and animal brains contrasted. Insanity. Magnetism. All shapes indicate character. How each faculty was discovered. Attestations. Objections. Organic quality. Mind shapes body. Form indicates the three temperaments — Vital, Motive, and Mental. Their combinations. Signs of character — complexion, eyes, beauty, walk, resemblance to animals, &c. Self-culture. Proportion the great law. Use strengthens. Self-knowledge. How to excite faculties, &c.

PART II. Health. Life's first pre-requisite. Amount attainable. Restorable. A duty. Sickness sinful. Life's functions. Vitality first. The Will cure. Respiration. Oxygen. Doubled by diaphragm breathing. The lungs. Ventilation. Blue veins. Stooping. Breathing propels the blood. The breath cure. Prevention and cure of consumption. Food. Appetite and smell should select it. Man omnivorous. Cookery. Unleavened bread. Fruits, sweets, pastry, milk, &c. Mastication, quantity, &c. The digestive process. Stomach and liver. Constipation, prolapsus, dyspepsia, how caused and cured, &c. Fluids. The blood. Drinks. Soft water. Tea, coffee, alcoholic, and malt liquor hankerings, and their cure. Tobacco. The heart, kidneys, glands, excretions, &c. The skin. Animal Heat. Fire, clothes, the feet. Colds, and their cure, &c. Sleep, its uses, times, promotion, &c. Bones, muscles, the exercise cure, &c. Nerves. Insanity, and its cure. Hydropathy, electropathy, coldpathy, sun and earth pathies, &c. Cure of asthma, inflammations, rheumatism, neuralgia, burns, wounds, tumors, &c. Healthy women. 21 Health rules. What is as practically important, throughout all the pursuits of life, as a good, sound constitution, that base of all terrestrial functions and enjoyments? What are riches and honors, what is even life itself, without health? What is the value of a robust *family* over a sickly one? And how *horrible* is premature death! Learn in Part II. how to secure the former, and avoid the latter.

PART III. The Self-Caring Propensities. Analysis, location, description, and culture of Acquisition, Secretion, Destruction, Force, Love, Parental Affection, Friendship, Inhabitiveness, Continuity, Caution, Ambition, Dignity, and Firmness, with observations as to economy, commerce, railroads, insurance, fortunes, policy, self-defence, polygamy, rearing children, patriotism, home and its improvement, the gravel wall and octagon modes of building, cheap cisterns, care of self, a good name, aristocracy, credit, self-respect, perseverance, &c.

PART IV. Man's Moral Nature proves a God, and immortality; and discloses a new *system* of moral philosophy, ethics, religion, and theology.

Did all that is come by chance? or exists there, in very truth, a God, the great Creator and Governor of this magnificent universe? And if aye, what of His attributes, government, works, worship, and the allegiance due from man to his Maker? That is, What is the true THE-OLOGY?

Is DEATH OUR LAST? or is man indeed immortal? And if so, what *of* that immortality? Are this life and that to come antagonistic? And if *so*, should we sacrifice the pleasures of this to those of that, or those of that for those of this? Or are both so interrelated that whatever promotes or curtails the pleasures of either, thereby similarly affects the other also? And if so, what life is best for us, both here and hereafter?

Is MAN NATURALLY DEPRAVED? And if he is, are there any antidotes, or even palliatives? Must he be born again? What of Faith, Prayer, Worship, Rites, Sin, Forgiveness, the Resurrection, &c.?

THESE and like problems, O man, which have puzzled the race throughout all ages, are among the most practically important mankind can ever ask or answer; because there impinge upon them eventualities so much farther reaching, and more momentous, than upon any others whatsoever, that it becomes us, as intelligent, self-interested beings, to obtain answers so absolutely reliable that we can well afford to live and DIE by them.

MAN'S MORAL FACULTIES solve these and all kindred problems scientifically and certainly; because they are adapted to, and put man in relation with, all the moral and religious principles and truths of the entire universe; so that, if there *is* a God, and if man is immortal, his moral faculties will be adapted to both; and if they are thus adapted, a God and an immortality certainly do exist. In short, Phrenology, in their analysis, unfolds all their relations and dependencies, together with all those ranges of truths they involve; and thus becomes a complete storehouse of moral and religious truths; besides unfolding a perfect SYSTEM of natural Theology. Please scan its following

SYNOPSIS. Man created with moral and religious *faculties*. These have their laws; which render Religion an exact, demonstrative natural science; and the same forever. Conflicting sects prove each other's erroneousness. This moral group, located above all else, does and should control human character and destiny; and joins, and should be exercised with reason. Our religious faculties are our teachers and text-book. Each of us should be our *own* priest and prophet.

MAN'S WORSHIPPING NATURE presupposes and proves that a God exists to *be* worshipped, just as his eating faculty implies food. All are born worshipful; therefore all are solemnly bound to worship. Devotion yields our richest pleasures, and sanctifies all our other enjoyments. All should study and obey God *in His works and natural laws*. Loving Him renders us like Him. Prayer benefits us, not Him. How it is answered. Religious creeds, rites, Sunday, &c. Sects accounted for. A new sect proposed. Analysis of the Divine attributes. The true Theology, &c.

SPIRITUALITY proves immortality, by adapting man to it; so does *progression*, general and individual; it does *Hope*, by expecting it. Age ripens, improves, and fits us for a spiritual existence. The origin of life, and even death itself, prove a life to come. This life merges into that. All our faculties here continue there, and have kindred objects. All we do and are here, affect us forever; just as our youthful conduct influences all after life. Memory there must recall every little event here. Life is a *system of causes* which produce eternal effects. Little things here cause great effects there. What causes here effect what results there.

"MINISTERING ANGELS." Communing with departed spirits. Forewarnings, visions, dreams, &c. How "special providences" are effected. Total depravity. Its origin, and obviation. Death, life's crowning blessing. A luxury to be craved, not evil to be dreaded. Kills all organic ills and vices. The Resurrection. Phrenology guides to a perfect life here and hereafter.

CONSCIENCE renders right and wrong inherent. All right self-rewarding; all wrong, self-punishing, by natural law, irrespective of faith, prayer, and persons. All suffering remedial, and will ultimately reform all. All evils produce good. Penitence due, implies forgiveness, stops further sin and suffering, and exalts morality. The Law of Love. Charity a duty. Blessed to give. The teachings of Phrenology harmonize with those of Christ. Summary application of this whole subject.

THIS PURELY SCIENTIFIC exposition of these and kindred subjects, from the constitution of the human mind, merits attention from all Christians, infidels, and savants. Is religion a myth to be ignored? Is Jehovah to be neglected? Are His works and worship, character and laws, an unheeded fleeting cloud? Is the study of your own moral constitution dry? No human investigations at all equal in utility and inherent interest, those of the moral constitution of the universe, of the fundamental principles of religion, and of the Divine attributes and commands. Learn in Part IV. that a Supreme Ruler over all really does exist, and the true theology, and that immortality is a veritable reality; that our "future state, is not hidden, not even beclouded; that as Moses from Pisgah's towering heights could discern Canaan's hills and vales," lakes and rivers, so we can diagnose, *scientifically*, "the land we're going to," even in detail.

PART V. — INTELLECT — Memory, reason, and their culture — analyzes, and shows how to cultivate, our senses and intellectual faculties; describes each in five different degrees of power, together with their combinations; shows how to conduct juvenile and self-education, musical, scientific, scholastic, conversational, &c.; and develop intellect, that highest department of man.

MEMORY is a most valuable possession. What rent could not lawyers, business men, scholars, everybody well afford to pay, to be enabled to recall and apply *all they ever knew!* How many daily losses, consequent on a poor memory, would a good one convert into gains! By disclosing Nature's true educational principles, parents, teachers, and individuals are here shown how to advance their own and children's mental culture many times faster than now.

REASON and *sense* are still more valuable; while learning, eloquence, and the other intellectual endowments are scarcely less so. In short, MIND controls matter. *Knowledge* is power. *Reason* is man's constitutional guide and governor in all things. Those alone may justly exalt who install sense as their ruler. *Mental discipline* is man's highest attainment; because it crowns all others. Teaching men the natural laws and the consequences of their obedience and infraction, enlists their very *self-interest* in leading right lives. Study, that you may discipline intellect and strengthen memory. What pleasures surpass those derived from reading and studying Nature, her laws and facts, philosophies and truths? Admeasure the value and pleasure of a strong and cultivated mind, over one dull and ignorant, and learn in Part V. how to realize them all.

THESE and kindred subjects, on man, go right home to your very life centre. One life alone is yours to enjoy, and improve. By all *its* value, do learn to make the *very* most out of *every single one* of its many powers and functions. Here, but *nowhere else*, can you find their analysis, with specific directions for their right exercise and culture, and warnings against their wrongs. *No other reading, no other* MEANS, will equally improve your entire being, mental and physical, now and forever. The vast range of human interests here discussed, entitles it to the patronage of all Phrenologists, Philanthropists, Philosophers, parents, and whoever would improve themselves.

BOTH WORKS CAN BE HAD after each lecture, at his rooms, and by remitting a $4 P. O. order, for each in muslin, $5 for leather, to O. S. Fowler, 514 Tremont Street, Boston, Mass. Agents wanted.

FOWLER ON SEXUAL SCIENCE AND RECUPERATION;

Or. MANHOOD, WOMANHOOD, and their Mutual Inter-relations; LOVE, its Laws, Power, &c.; SELECTION, or Mutual adaptation; COURTSHIP, or Love Making; MARRIED LIFE made happy; REPRODUCTION, and Progenal Endowment, or Paternity, Maternity, Bearing, Nursing, and Rearing Children; PUBERTY, Girlhood, &c.; SEXUAL ALIMENTS restored, and Female beauty perpetuated, &c., as taught by Phrenology.

A RIGHT MALE AND FEMALE LIFE constitutes the master problem, as yet unsolved, of every human being. "SEXUAL SCIENCE" expounds this problem, and thereby utters a divine mandate to lovers, the married, and prospective parents, to learn and fulfil Nature's sexual laws. No other knowledge is equally important, because no other duties are as imperious; nor is ignorance of any other equally fatal. Reader, would not such knowledge have converted your own present sexual or marital sufferings into enjoyments? By expounding and applying Nature's sexual ordinances to these subjects, among many other kindred ones which it discusses, does it not go right home to the heads and hearts of all who have either?

PART I. MANHOOD and WOMANHOOD, or the constituent elements of male and female perfection. Dignity and utility of sexual knowledge. Reproduction. Nature's paramount work. "Each after its own kind," throughout every minute particular of body, mind, instinct, everything. SEXUAL ATTRACTION its means. Love confers procreative capacity and conjugal talents. Is located at the seat of physical life, and apex of every mental organ. Is in sympathetic *rapport* with every iota of all, that it may transmit all to progeny. Gender wields supreme control over the voice, walk, dance, beauty, complexion, eyes, courage, talents, temper, spirits, morals, happiness, and every element of body and mind. A right sexual state impairs, a wrong crushes, the entire being. Hybrids show what parts are derived from the male, and what from the female. What each sex likes and dislikes in the other, and why. Woman loves originality, power, firmness, courage, passion, &c., in man, but dislikes their converse, because the male originates life, and most things human, and confers these elements on offspring; while man loves purity, exquisiteness, affection, maternal love, piety, taste, prudence, ton, &c., in woman, because mothers stamp these attributes on children. How all women can obtain much more than their "rights." How ladies and gentlemen should treat each other. Signs of a strong and weak, healthy and impaired, sexuality. Male and female forms contrasted and criticised. Only maternal excellences create female beauty. Why men admire the female bust. Analysis of the fashions, &c.

PART II. LOVE. Its analysis in all its different aspects. Its magic power over the entire being. Its right state improves, but wrong impairs, the form, walk, countenance, muscles, circulation, health, longevity, expression, tones, laugh, manners, mental faculties, energy, industry, ambition, self-trust, morals, hopes, worship, kindness, taste, wit, memory, music, language, sense, agreeableness, — every function of body and mind. Makes or breaks all. Is life's "master passion." MARRIAGE is its natural sphere. Pairing innate. Mating before twenty-one for women, and twenty-three for men, a solemn duty. Its poor substitute. Old-bachelorism, old-maidism, and their excuses. One love *vs.* free-love. Marriage creates families and homes. All sexual vices originate in disappointed or perverted love. How to moralize young men, heal "broken hearts," redeem sexual sinners, restore one's own self to purity, &c. Seducers most accursed. Self-abuse, its prevalence, and terrible effects. It exhausts, inflames, unsexes and destroys the entire body and mind. Is as sinful as fornication. Its prevention by knowledge, conscience, commingling of the sexes, &c. Boys and girls, mixed schools, fathers and daughters, mothers and sons, brothers and sisters, ladies and gentlemen, public and parlor amusements, &c.

PART III. Selection, or the natural *laws* of sexual attraction and repulsion. Founding a family. Nature's true *time* to mate and wed. Similar and different ages. A right choice life's casting die. Mutual rights of parents and children in their own and each other's selection. *Self* the final umpire. Courtship's first stage. *General* marital pre-requisites. Healthy *vs.* invalid, and housekeeping *vs.* fashionable, consorts. Wealth *vs.* worth. A poor *man vs.* a rich *thing.* Habits, temperance, sexuality, &c. Marrying cousins. What traits each requires in the other. Superb offspring the determining condition. When and why those similar, and dissimilar, should, and must not, marry. Combining the greatest aggregate number and amount of excellences. How to find one thus adapted. Phrenology aids a right choice. Intuition. The proposal, acceptance, and vow. Parental consent, relatives, elopements, dismissals, breaches of promise, &c.

PART IV. Courtship, its fatal errors and right management. None should love until engaged. Loving is marrying. Love-spats. Flirtations. Liberties. Presents. Disclosing faults. Day *vs.* night, and Sunday evening courtships. Sudden loves. Right courtship. Its first pre-requisite. Duration, &c.

PART V. Married Life. Establishing a perfect love. The wedding, honey-moon, honey-marriage, first year, &c. Love-making rules. 1. Be perfect gentlemen and ladies. 2. Mould and be moulded. 3. Co-operate in all things. 4. Promote each other's happiness. 5. Nurture each other's affections more than during courtship. Discords, their amount, causes, and obviation. Toleration. Burying old bones, &c. Divorces. When, and when not, allowable, &c.

PART VI. Reproduction. The ultimate end of everything sexual. Intercourse its only means. Its governing condition. All temporary parental states transmitted. Platonic love the great pre-requisite. It creates the *mind.* Love and intercourse mutual concomitants; therefore marrying one while loving another is double adultery. Love and the sexual organism in sympathetic *rapport.* Female passion necessary. It inspires man. Why intercourse without it becomes insipid, injurious, and most infuriating. Its promotion by health, love, &c. Amorous husbands and passive wives, reproved. Woman may control her own person. Conjugal rapes. What creative states promote, and what prevent, parental pleasure, and progenal endowment. Promiscuosity. Frequency. Advice to those just married. Mutual adaptation of the sexes. The life-germ; its creation, progress, wants, and their supply. The female creative office. The womb, and its appendages. The ovum, and its fecundation. Pre-determining the sex. Twins and triplets. Promoting and preventing conception. Barrenness, onanism. &c., &c.

PART VII. Maternity, and childbirth. The effects, on offspring, of different antenatal states. Ishmael, Samuel, Samson, Christ. James I., Bonaparte, the giant maniac, &c. Why children should be loved before their birth, bad tempered ones be pitied. &c. Diseases, marks, hydrocephalus, &c., their causes, and preventives. Intercourse during pregnancy. Maternal vitality, sleep, food, breathing, tight-lacing, exercise, fear, fortitude, inanity, culture, &c. The first six months, and last three.

How to render children natural divines, poets, scholars, thinkers, business men, artists, &c. Maternity should take precedence. Pregnancy healthy. Rendering childbirth easy by health, muscular culture, resolution, water treatment, &c. What forms may, and must not, intermarry. Drugging, bleeding, milk sickness, prompting lactation, &c.

PART VIII. CHILD-BEARING, its laws and details. Modern education empirical. Value of infants. Their nursing, " complaints," teething, worms, scarlet fever, crying, weaning, diet, habits, sleep, ablution, clothing, going barefoot, schooling, &c. Their nutritive, muscular, and growing epochs. Play, and playmates. Precocity. Governing them by moral suasion *vs.* the rod. Directing will, not crushing it. Example *vs.* precept. Patience *vs.* scolding. Love the mother's magic wand, &c.

PART IX. SEXUAL RECUPERATION. The amount, causes, and cures of sexual disorders. Right and wrong love states. Continence. Seminal losses. Prematurity. Impotence. Aversions. Health habits, exercise, &c. Local applications of water, electricity, &c. Female complaints, their causes and cures. Girlhood. Right and wrong merging into womanhood. Sexual inertia. Female ruination, misnamed education. Abortion, and sexual frauds. Prolapsus, its effects and cure. Visceral manipulations. Fluor albus. Miscarriages, and their prevention. Menstruation, its office and promotion. Surplus fat, and labored breathing, their causes and cures. Barrenness, its causes and obviation. Female beauty. Its conditions. A full bust. How lost, and regained. Rules for promoting sexual vigor. Concluding appeal.

Reader, can you get a tithe as much life-long value with the same money, as by procuring this volume? YOUR SOCIAL and sexual nature literally controls your entire being. Three great ends impinge upon it: sexual perfection, conjugal happiness, and perfect children. What other human good equals either? Let all tremble in view of the magic power wielded by gender over the whole being, and here learn how to escape all its evils, and enjoy all its benefits. In the language of its preface, —

" Reader, go none of *these* subjects right *home* to the very heart's *core* of your inner *life !* Have you no masculine or feminine nature to study, direct, nurture, enjoy, or recuperate? Have you neither conjugal mate, nor any tender yearnings for some loved one to inspire hope, incite to effort, share life's joys and sorrows, and tread with you the pathways of earth and heaven? Have you no children, and no wish for any, to inherit your mentality and physiology, as well as patrimony? to do and to care for? and to care and do for you? to close your eyes in death? and after it to repeat your virtues? In fine, are you listless, aimless, forlorn driftwood, left by the ebbing current of time, sinking and decaying in the mire of inanity; none caring for you, and you for none? For if not all these, and much more, then should the *subject-matter* of this volume stir your soul to its innermost depths, and sweep whatever life chords remain unpalsied within you. Nothing else lies quite as near the focal centre of human existence as do our affections; and this treatise will show all how to derive from them the most enjoyment possible, with the least suffering. It assumes all the dignities and immunities of a thoroughly SCIENTIFIC and purely philosophical treatise on the whole subject of man's domestic, social, and sexual constitution and relations. Where have they ever before been discussed thus *collectively ?*"

ALL O. S. FOWLER'S WORKS AND WRITINGS are embraced in these two volumes, which, together, constitute a complete exposition of human nature, physical, passional, sentimental, affectional, moral, and intellectual.

IT CAN BE HAD after each lecture, or at his rooms, and also by remitting a $4 P. O. order, for muslin, and $5 for leather, to O. S. Fowler, 514 Tremont Street, Boston, Mass. Published only by subscription. Agents wanted.

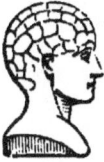